Conference of Theoretical Physics and Nonlinear Phenomena (CTPNP 2014)

From Universe to String's Scale

Journal of Physics: Conference Series Volume 539

Surakarta, Indonesia
15 February 2014

Editors:

Utama Alan Deta M. A. Cari

ISBN: 978-1-63439-464-2
ISSN: 1742-6588

Conference of Theoretical Physics and Nonlinear Phenomena (CTPNP 2014)

From Universe to String's Scale

Journal of Physics: Conference Series Volume 539

Surakarta, Indonesia
15 February 2014

ISBN: 978-1-63439-464-2
ISSN: 1742-6588

IOP Institute of Physics

Printed from e-media with permission by:

Curran Associates, Inc.
57 Morehouse Lane
Red Hook, NY 12571

Some format issues inherent in the e-media version may also appear in this print version.

Printed by Curran Associates, Inc. (2014)

For permission requests, please contact the Institute of Physics
at the address below.

Institute of Physics
Dirac House, Temple Back
Bristol BS1 6BE UK

Phone: 44 1 17 929 7481
Fax: 44 1 17 920 0979

techtracking@iop.org

Additional copies of this publication are available from:

Curran Associates, Inc.
57 Morehouse Lane
Red Hook, NY 12571 USA
Phone: 845-758-0400
Fax: 845-758-2634
Email: curran@proceedings.com
Web: www.proceedings.com

Table of contents

Volume 539

Conference of Theoretical Physics and Nonlinear Phenomena (CTPNP) 2014: "From Universe to String's Scale"
15 February 2014, Surakarta, Indonesia

Accepted papers received: 22 August 2014
Published online: 13 October 2014

Preface

Papers

Plenary Session

Parallel 1: Astronomy and Astrophysics

Parallel 2: High Energy Physics

Parallel 3: Mathematical Physics

Parallel 4: Quantum Physics

Parallel 5: Non–Linear Physics

Journal of Physics: Conference Series **539** (2014) 011001 doi:10.1088/1742-6596/539/1/011001

Conference of Theoretical Physics and Nonlinear Phenomena (CTPNP) 2014
"From Universe to String's Scale"
http://situs.opi.lipi.go.id/ctpnp2014/

Introduction

A. Background

Theoretical physics is the first step for the development of science and technology. For more than 100 years it has delivered new and sophisticated discoveries which have changed human views of their surroundings and universe. Theoretical physics has also revealed that the governing law in our universe is not deterministic, and it is undoubtedly the foundation of our modern civilization.

Contrary to its importance, research in theoretical physics is not well advanced in some developing countries such as Indonesia. This workshop provides the formal meeting in Indonesia devoted to the field of theoretical physics and is organized to cover all subjects of theoretical physics as well as nonlinear phenomena in order to create a gathering place for the theorists in Indonesia and surrounding countries, to motivate young physicists to keep doing active researches in the field and to encourage constructive communication among the community members.

Following the success of the tenth previous meetings in this conference series, the eleventh conference was held in Sebelas Maret University (UNS), Surakarta, Indonesia on 15 February 2014. In addition, the conference was proceeded by School of Advance Physics at Gadjah Mada University (UGM), Yogyakarta, on 16-17 February 2014.

B. Objectives

The conference is expected to provide distinguished experts and students from various research fields of theoretical physics and nonlinear phenomena in Indonesia as well as from other continents the opportunities to present their works and to enhance contacts among them.

C. Invited Speakers

The Scientific Program of CTPNP 2014 was supposed to include plenary talks given by the following invited speakers:

1. Prof. Ernest Ma (UC Riverside, California)*
Unification of Matter and Dark Matter with Radiative Neutrino Mass

2. Martin Spinrath, Ph.D. (SISSA, Trieste, Italy)*
GUT Scale Fermion Mass Ratios

3. Anto Sulaksono, Ph.D. (UI, Indonesia)
Hyperonsin Anisotropic Neutron Star

4. Suharyo Sumowidagdo, Ph.D. (CERN, Swiss and LIPI, Indonesia)*
LHC Accelerator and Experiments at CERN

5. Agus Purwanto, D.Sc. (ITS, Indonesia)*
Quantum Heat Engine

6. Tri L. Astraatmadja, Ph.D. (MPIA, Germany)
Neutrinos from GRBs and their Detection with the ANTARES Neutrino Telescope

7. Suparmi, Ph.D. (UNS, Indonesia)
Exact Solution of Dirac Equation for q-Deformed Trigonometric Scarf Potential with q-Deformed Trigonometric Tensor Coupling Potential for Spin and Pseudospin Symmetries Using Romanovski Polynomial

Journal of Physics: Conference Series **539** (2014) 011001 doi:10.1088/1742-6596/539/1/011001

*could not attend the conference because of Kelud Mountain's Eruption.
It is also planned that in each of the parallel sessions there will be a lead talk given by theyoung scientists. The presented and selected manuscript of CTPNP 2014 are published in Journal of Physics: Conference Series (indexed by SCOPUS).

D. Participants

Researchers and students of theoretical physics and nonlinear phenomena from Indonesia and other country (especially in South East Asia) as well as from other parts of the world were kindly invited to participate in this conference.

E. Expert Referees

To ensure the quality of published article, the manuscript will be peer reviewed by anexpert referees, that is:

1. Prof. Pramudita Anggraita
Atomic Nuclear Energy Agency (BATAN), Indonesia

2. Martin Spinrath, Ph.D.
Institut fur TheoretischeTeilchenphysik, Karlsruhe Institute of Technology (KIT), Germany

3. Dr. L.T. Handoko
Indonesian Institute of Science (LIPI), Jakarta

4. Anto Sulaksono, Ph.D.
Physics Department, University of Indonesia (UI), Indonesia

5. Suparmi, Ph.D.
Physics Department, Sebelas Maret University (UNS), Indonesia

Journal of Physics: Conference Series **539** (2014) 011002 doi:10.1088/1742-6596/539/1/011002

Conference on Theoretical Physics and Non-Linear Phenomena (CTPNP) 2014

Advisory Board

Organizer

A. Bangsawang

Hasanudin University (UNHAS), Makassar

A. Damanik

Sanata Dharma University (USD), Yogyakarta

A.N. Atmaja

Indonesian Institute of Science (LIPI), Serpong

A. Purwanto

Sepuluh Nopember Institute of Technology (ITS), Surabaya

A. Sulaiman

Agency for The Assessment and Application of Technology (BPPT), Jakarta

A.Y. Wardaya

Diponegoro University (UNDIP), Semarang

B. E. Gunara

Bandung Institute of Technology (ITB), Bandung

E. Latifah

State University of Malang (UM), Malang

F.P. Zen

Bandung Institute of Technology (ITB), Bandung

F. Rusyidi

Airlangga University (UNAIR), Surabaya

H. Alatas

Bogor Agricultural University (IPB), Bogor

I. Fachrudin

University of Indonesia (UI), Depok

L. Rohman

State University of Jember (UNEJ), Jember

L.T. Handoko

Indonesian Institute of Science (LIPI), Jakarta

M. Satriawan

Gajah Mada University (UGM), Yogyakarta

P. Anggraita

National Nuclear Energy Agency (BATAN), Yogyakarta

P. W. Premadi

Bandung Institute of Technology (ITB), Bandung

Roniyus M.S.

Lampung University (UNILA), Lampung

S. Feranie

Indonesia University of Education (UPI), Bandung

Journal of Physics: Conference Series **539** (2014) 011002 doi:10.1088/1742-6596/539/1/011002

Journal of Physics: Conference Series **539** (2014) 011002 doi:10.1088/1742-6596/539/1/011002

Conference on Theoretical Physics and Non-Linear Phenomena (CTPNP) 2014
Local Organizer on Sebelas Maret University (UNS)

Advisor : Drs. Cari, M.A., M.Sc., Ph.D.

Chairman : Dra. Suparmi, M.A., Ph.D.

Vice Chairman : 1. Dr. Yoventina Iriani, M.Si.
2. Utama Alan Deta, M.Pd., M.Si.

Secretary : 1. Dr. Eng. Budi Purnama, M.Si.
2. ST. Nurul Fitriani, S.Pd.

Treasurer : 1. Ahmad Marzuki, S.Si., Ph.D.
2. Alfi Zaidah, S.Pd.

➢ **Event Organizer**
1. Hardani, S.Pd. (Coordinator)
2. Hendra Darmaja, S.Pd.
3. Luqman Hakim, S.Si.
4. Supriyanto, S.Pd.
5. Purwo Suyono, S.Pd.

➢ **Registration**
1. Suwarni, S.Si. (Coordinator)
2. Alpiana Hidayatulloh, S.Pd.
3. Kholida Ismatulloh, S.Pd.
4. Tri Jayanti, S.Si.

➢ **Master of Ceremony**
Amalia Puji Winarni

➢ **Logistics, Publications, and Documentation**
1. Moh. Iman Darmawan, S.Pd. (Coordinator)
2. Fahrudin, S.Si.

➢ **Moderator**
1. Dr.Eng. Nuryani, M.Si.
2. Khairudin, S.Si., Ph.D.
3. Ahmad Marzuki, S.Si., Ph.D.
4. Dr.Eng. Budi Purnama, M.Si.

- **Proceeding Editor**

    1. Utama Alan Deta, M.Pd., M.Si.

    2. Drs. Cari, M.A., M.Sc., Ph.D.

Conference on Theoretical Physics and Non-Linear Phenomena (CTPNP) 2014
List of Participant

Speakers

1. A. Hermanto
2. A. Suparmi
3. Ardian Nata Atmaja
4. Asan Damanik
5. Budi Purnama
6. C. Cari
7. C. Monstein
8. D. Andra
9. D.S. Palupi
10. E. Suhendi
11. E. Tenderlilin
12. F.A. Noor
13. Gusrizal
14. Julio
15. Khairurrijal
16. Luqman H.
17. M.B. Ibrahim
18. M.F. Rosyid
19. Miftachul Hadi
20. N. Kurniasih
21. N.A. Amran
22. N.N.M. Shariff
23. N.S. Arifin
24. R. Syariati
25. Roslan Umar
26. S. Wahyuni
27. S.K. Zulkeple
28. Sudi Mungkasi
29. Supriyanto
30. U.A. Deta
31. W. Sadnawanto
32. W.N.A. Wan Zulkifli
33. W.S. Putro
34. W.S.B. Dwandaru
35. Wayan Suparta
36. Yunita Umniyati
37. Z.A. Ibrahim
38. Z.S. Hamidi
39. Zainol Abidin Ibrahim
40. Zamri Zainal Abidin

Participants

1. Achmad Prayogi
2. Ade Kurniawan
3. Ahmad Marzuki
4. Alfi Zaidah
5. Alpiana Hidayatulloh
6. Amalia Puji Winarni
7. Annisa Fitri
8. Asmiati

Journal of Physics: Conference Series **539** (2014) 011003 doi:10.1088/1742-6596/539/1/011003

9. Dian Wahyu Nur Ivanty
10. Elida Lailiya Istiqomah
11. Fahrudin
12. Hardani
13. Hasan Bisri
14. Hendra Darmaja
15. Ibrahim Hasan
16. Ilham Prasetyo
17. Khairi Trisnayadi
18. Khairudin
19. Kholida Ismatulloh
20. Moh. Iman Darmawan
21. Muhammad Ali
22. Muhammad Iqbal
23. Muhammad Syaifudin
24. Novi Ayu Kristiana Dewi
25. Nuryani
26. Omar Moftah Mokhtar Moyouf
27. Pramudita Anggraita
28. Purwo Suyono
29. Rodi
30. ST. Nurul Fitriani
31. Suwarni
32. Tri Jayanti
33. Yoventina Iriani
34. Yudi Pramono Pawiro
35. Yulia Dewi Puspita Sari
36. Yusuf Nurochman

There were another speakers and participants who registered to the CTPNP 2014. Because of Kelud Mountain's Eruption, they could not attend this conference.

Journal of Physics: Conference Series **539** (2014) 011004 doi:10.1088/1742-6596/539/1/011004

Conference on Theoretical Physics and Non-Linear Phenomena (CTPNP) 2014 Acknowledgement

The editor and organizer of the Conference on Theoretical Physics and Non-Linear Phenomena (CTPNP) 2014 wish to acknowledge the keynote speakers for their presentation on CTPNP 2014. The organizers also wish to acknowledge publicly the valuable services provided by the reviewers. On behalf of the editors, organizers, authors and readers of this Conference, we wish to thank the keynote speakers and the reviewers for their time, hard work, and dedication to this Conference. Without their services, the editors could not maintain the high standards of Theoretical Physics and Non-Linear Phenomena Research.

The organizers wish to acknowledge Prof. Pramudita Anggraita, Dr. L.T. Handoko, Dr. Husin Alatas, and Dr. Mirza Satriawan for the discussion, suggestion, and cooperation to organize the keynote speakers of this Conference. The organizers also wish to acknowledge for speakers and participants who attend this seminar although there is a big natural disaster (eruption of Kelud Mountain) one day before the conference. We also acknowledge registered speakers and participant who cannot attend this conference due to the Kelud Mountain's eruption. Many thanks given for all persons who help and support this conference.

The organizer wish to apology to the speakers who cannot publish his/her paper in this Conference Proceeding. Our apology also given to all participants for all shortcomings in this conference. See you in the next CTPNP 2015 at the Indonesian Institute of Science (LIPI).

Surakarta, March 17, 2014

Local Organizer of CTPNP 2014
Physics Department
Sebelas Maret University

Conference Photos

Journal of Physics: Conference Series **539** (2014) 011005 doi:10.1088/1742-6596/539/1/011005

Journal of Physics: Conference Series **539** (2014) 011005 doi:10.1088/1742-6596/539/1/011005

WELCOME TO
CONFERENCE ON THEORETICAL PHYSICS
AND NONLINEAR PHENOMENA 2014

Peer review statement

All papers published in this volume of *Journal of Physics: Conference Series* have been peer reviewed through processes administered by the proceedings Editors. Reviews were conducted by expert referees to the professional and scientific standards expected of a proceedings journal published by IOP Publishing.

Journal of Physics: Conference Series **539** (2014) 012001 doi:10.1088/1742-6596/539/1/012001

Unification of Matter and Dark Matter with Radiative Neutrino Mass

Ernest Ma

Department of Physics and Astronomy, University of California, Riverside, California 92521, USA

E-mail: ma@phyun8.ucr.edu

Abstract. Neutrino mass may be linked to dark matter through a well-studied radiative (scotogenic) mechanism. The new particles required are just right for the extension of the well-known $SU(5)$ unification of quarks and leptons to include dark matter. Two examples are discussed: $SU(6)$ which incorporates a discrete Z_2 symmetry for dark matter, and $SU(7)$ which allows for a gauge $U(1)_D$ symmetry. In either case, just as the proton is not absolutely stable within the context of $SU(5)$, dark matter is also not absolutely stable within $SU(6)$ or $SU(7)$.

1. Introduction

Physics beyond the standard model (SM) of quarks and leptons should include neutrino mass and dark matter (DM). Are they related? I proposed [1] in 2006 that neutrino mass is due to the existence of dark matter (scotogenic = caused by darkness). I will discuss the prototype model based on the discrete symmetry Z_2, and its recent variation [2] with a gauge $U(1)_D$ symmetry. Just as quarks and leptons may be unified in $SU(5)$, I will show how matter and dark matter may be unified in $SU(6)$ and $SU(7)$.

2. Neutrino Mass: Six Generic Mechanisms

In 1979, it was first pointed out by Weinberg [3] that a unique dimension-five operator exists for Majorana neutrino mass in the SM, assuming that new physics occurs at a mass scale higher than all the masses of the SM. This effective interaction

$$\mathcal{L}_5 = -\frac{f_{\alpha\beta}}{2\Lambda}(\nu_\alpha\phi^0 - l_\alpha\phi^+)(\nu_\beta\phi^0 - l_\beta\phi^+) \tag{1}$$

became the starting point of most models of neutrino mass. In 1998, I discussed [4] for the first time together all three tree-level realizations of this operator, as well as three possible one-loop (one-particle-irreducible) realizations. To summarize, the three tree-level realizations are mediated by (I) a heavy neutral singlet Majorana fermion N, (II) a heavy scalar triplet (ξ^{++}, ξ^+, ξ^0), and (III) a heavy Majorana fermion triplet $(\Sigma^+, \Sigma^0, \Sigma^-)$. Ever since, these three mechanisms have been referred to as Type I, Type II, and Type III seesaw respectively.

As for the three generic one-loop (one-particle-irreducible) realizations, the loop is assumed to consist of a scalar line and a fermion line linking ν_α with ν_β. The three realizations are then classified according to (IV) one ϕ^0 attached to the scalar line and one ϕ^0 attached to the fermion line, (V) both attached to the scalar line, and (VI) both attached to the fermion line.

3. Scotogenic Neutrino Mass

In 2006, I proposed [1] a specific realization of (V) using a neutral fermion singlet N per family which is odd under an imposed discrete Z_2 symmetry. The interaction $(\nu\phi^0 - l\phi^+)N$ is then forbidden and ν does not combine with N in the usual way to obtain a Dirac mass, although a Majorana mass for N itself is allowed. To generate a nonzero neutrino mass, I assumed a second scalar doublet (η^+, η^0) which is also odd under Z_2, so that the interaction $(\nu\eta^0 - l\eta^+)N$ is allowed, but Z_2 is assumed to be unbroken (for dark matter) so η^0 has zero vacuum expectation value and again there is no Dirac neutrino mass. However, the quartic scalar term $(\lambda_5/2)(\Phi^\dagger\eta)^2 + H.c.$ is allowed, and the one-loop diagram (V) exists for a radiatively generated Majorana neutrino mass, as shown in Fig. 1. The one-loop integral is exactly calculated as

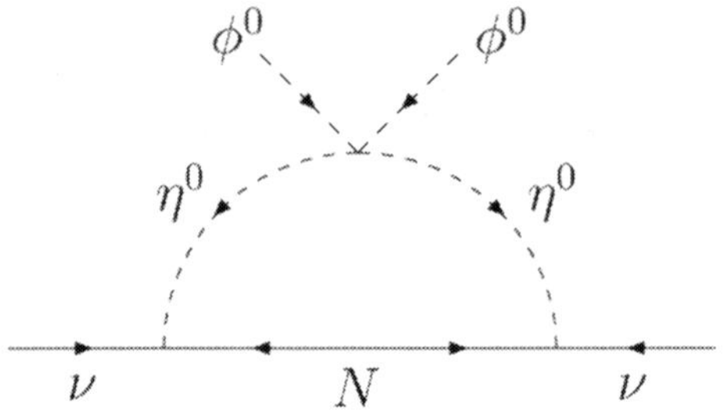

Figure 1. Scotogenic neutrino mass with Z_2.

$$(\mathcal{M}_\nu)_{\alpha\beta} = \sum_i \frac{h_{\alpha i} h_{\beta i}}{16\pi^2}[f(M_i^2/m_R^2) - f(M_i^2/m_I^2)], \tag{2}$$

where $f(x) = -\ln x/(1-x)$. Let $m_R^2 - m_I^2 = 2\lambda_5 v^2 << m_0^2 = (m_R^2 + m_I^2)/2$, then

$$(\mathcal{M}_\nu)_{\alpha\beta} = \sum_i \frac{h_{\alpha i} h_{\beta i}\lambda_5 v^2}{8\pi^2 M_i}\left(\frac{x_i}{1-x_i}\right)\left[1 + \frac{x_i \ln x_i}{1-x_i}\right], \tag{3}$$

where $x_i = M_i^2/m_0^2$.

4. New Scotogenic $U(1)_D$ Model

The notion that dark matter may have its own interaction, not shared by ordinary matter, has become increasingly relevant as a means of understanding the details of the dark-matter density profile of (dwarf) galaxies, i.e. the cusp-core discrepancy. To support a gauge $U(1)_D$ symmetry [2], two scalar doublets $(\eta_{1,2}^+, \eta_{1,2}^0)$ are used together with three Dirac singlet neutral fermions $N_{1,2,3}$ which transform as +1 together with η_1, whereas η_2 transforms as −1. The allowed interaction $(\Phi^\dagger\eta_1)(\Phi^\dagger\eta_2) + H.c.$ results in two neutral mass eigenstates

$$\begin{pmatrix} \chi_1 \\ \chi_2 \end{pmatrix} = \begin{pmatrix} \cos\theta & -\sin\theta \\ \sin\theta & \cos\theta \end{pmatrix} \begin{pmatrix} \eta_1^0 \\ \bar{\eta}_2^0 \end{pmatrix}. \tag{4}$$

Hence a one-loop neutrino mass may be generated as shown in Fig. 2 without breaking the gauge $U(1)_D$ symmetry. In that case, a massless dark photon would appear.

There are two other alternatives. A scalar $\zeta \sim 2$ under $U(1)_D$ may be used to break it to Z_2. If $\zeta \sim 3$, then $U(1)_D$ becomes a global symmetry. In either case, there will be a massive dark photon and a massive dark Higgs particle.

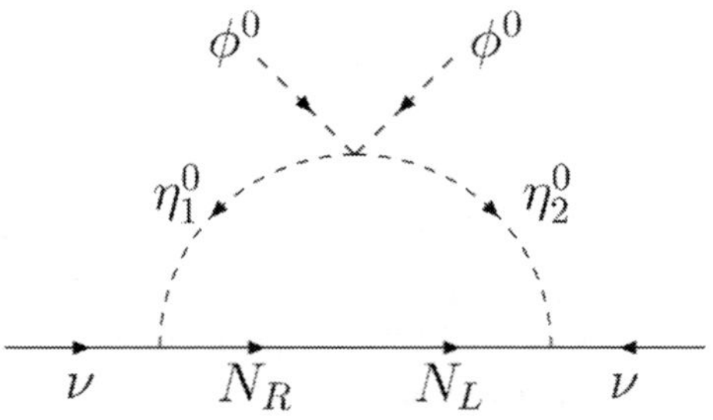

Figure 2. Scotogenic neutrino mass with $U(1)_D$.

5. $SU(6)$ Unification

To unify matter and dark matter, the well-known $SU(5)$ grand unification [5] of quarks and leptons may be simply extended [6] to $SU(6)$. The fundamental $\underline{5}^*_F = (d^c, d^c, d^c, e, \nu)$ of $SU(5)$ is extended to the $\underline{6}^*_F = (d^c, d^c, d^c, e, \nu, N)$ of $SU(6)$. The $\underline{10}_F$ of $SU(5)$ is extended by a heavy $\underline{5}_F$ to form a $\underline{15}_F$ of $SU(6)$. Together with the similar extensions of the scalar multiplets, the $SU(5)$ Yukawa terms

$$\underline{5}^*_F \times \underline{10}_F \times \underline{5}^*_S, \quad \underline{10}_F \times \underline{10}_F \times \underline{5}_S, \tag{5}$$

are extended to

$$\underline{6}^*_F \times \underline{15}_F \times \underline{6}^*_S, \quad \underline{15}_F \times \underline{15}_F \times \underline{15}_S. \tag{6}$$

Hence two different Higgs doublets are needed for quark and lepton masses.

Whereas $\underline{5}^*_F + \underline{10}_F$ is anomaly-free in $SU(5)$, the corresponding combination in $SU(6)$ is $\underline{6}^*_F + \underline{6}^*_F + \underline{15}_F$. The $\underline{5}^*_F$ contained in the extra $\underline{6}^*_F$ is heavy and pairs up with the $\underline{5}_F$ contained in the $\underline{15}_F$ through the $\underline{1}_S$ of $\underline{6}^*_S$.

Consider now the $SU(6)$ scalar multiplet $\underline{21}_S$. It decomposes into $\underline{15}_S + \underline{5}_S + \underline{1}_S$ of $SU(5)$. It has then the second scalar doublet (η^+, η^0) and the interactions

$$\underline{6}^*_F \times \underline{6}^*_F \times \underline{21}_S, \quad \underline{15}^*_S \times \underline{15}^*_S \times \underline{21}_S \times \underline{21}_S. \tag{7}$$

Hence N gets a Majorana mass through the $\underline{1}_S$ of $\underline{21}_S$. The scotogenic interaction $(\nu\eta^0 - e\eta^+)N$ is now possible as well as the quartic $(\Phi^\dagger\eta)^2$ interaction. These three terms support a Z_2 symmetry as required for Fig. 3, but it is not absolute. Just as the proton is unstable at the scale of quark-lepton unification, dark matter is expected to be unstable at a similar scale. Matter and dark matter are thus unified through scotogenic neutrino mass.

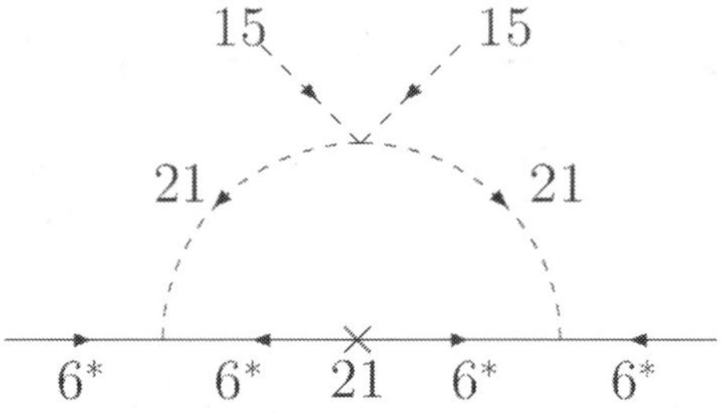

Figure 3. Scotogenic neutrino mass in $SU(6)$.

The heavy color triplet gauge bosons contained in the adjoint $\underline{24}_V$ of $SU(5)$ mediate proton decay. The extra heavy color triplet gauge bosons contained in the adjoint $\underline{35}_V$ of $SU(6)$ will connect N to the quarks, resulting in $N \to p\pi^-, n\pi^0$. Another possibility is the mixing of the heavy scalar color triplet $\zeta^{-1/3}$ in $\underline{21}_S$ with its counterpart $\xi^{-1/3}$ in $\underline{15}_S$ through the adjoint $\underline{35}_S$ of $SU(6)$. The dark Z_2 is thus broken at a high scale, just as baryon number is supposed to be broken in grand unified theories.

In the breaking of $SU(6)$ through the adjoint $\underline{35}_S$, let the 6×6 matrix representing $\underline{35}_S$ develop diagonal vacuum expectation values proportional to $(1,1,1,0,0,-3)$; then the residual symmetry is $SU(3)_C \times SU(2)_L \times U(1)_Y \times U(1)_N$. The dark $U(1)_N$ is broken by $\underline{21}_S$ at a lower energy scale for N to acquire a Majorana mass. This choice of $\underline{35}_S$ breaking mixes ζ with ξ, but not η with Φ. It is only possible here because of $SU(6)$. For $SU(5) \to SU(3)_C \times SU(2)_L \times U(1)_Y$, it must be along the $(1,1,1,-3/2,-3/2)$ direction.

6. $SU(7)$ Unification

In the decomposition $SU(5) \times SU(2)_N$, the $\underline{7}^*_F$ of $SU(7)$ contains an extra $SU(2)_N$ doublet (N_1, N_2). The $\underline{10}_F$ of $SU(5)$ is now extended to $\underline{21}_F$ of $SU(7)$ containing two $SU(5)$ $\underline{5}_F$s transforming as a doublet under $SU(2)_N$. The anomaly-free combination is $\underline{7}^*_F + \underline{7}^*_F + \underline{7}^*_F + \underline{21}_F$. The corresponding Yukawa couplings are

$$\underline{7}^*_F \times \underline{21}_F \times \underline{7}^*_S, \quad \underline{21}_F \times \underline{21}_F \times \underline{35}_S. \tag{8}$$

The $\underline{28}_S$ of $SU(7)$ now contains a bidoublet

$$\begin{pmatrix} \eta_1^+ & \eta_2^+ \\ \eta_1^0 & \eta_2^0 \end{pmatrix}, \tag{9}$$

where $SU(2)_L$ applies vertically and $SU(2)_N$ applies horizontally. To complete the scotogenic loop as shown in Fig. 4 with $U(1)_D$ as the diagonal subgroup of $SU(2)_N$, $\underline{28}_S$ is added with the interactions

$$\underline{7}^*_F \times \underline{7}^*_F \times \underline{28}_S, \quad \underline{21}^*_S \times \underline{21}^*_S \times \underline{28}_S \times \underline{28}_S. \tag{10}$$

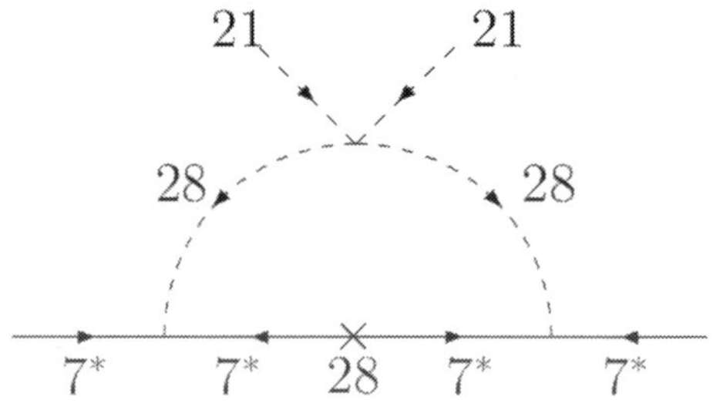

Figure 4. Scotogenic neutrino mass in $SU(7)$.

7. Conclusion

The evidence of dark matter signals a new class of particles at the TeV scale, which may manifest themselves indirectly through loop effects. They may be responsible for radiative neutrino mass and facilitate the unification of matter and dark matter.

Two prototype models based on $SU(6)$ and $SU(7)$ are presented. The former applies naturally to the original scotogenic model of neutrino mass with Z_2 symmetry. The latter fits well with a recent $U(1)_D$ extension.

Acknowledgments
I thank the organizers of CTPNP for their great hospitality and a stimulating conference. This work is supported in part by the U. S. Department of Energy under Grant No. DE-SC0008541.

References

[1] Ma E 2006 *Phys. Rev. D* **73** 077301
[2] Ma E, Picek I and Radovcic B 2013 *Phys. Lett. B* **726** 744
[3] Weinberg S 1979 *Phys. Rev. Lett.* **43** 1566
[4] Ma E 1998 *Phys. Rev. Lett.* **81** 1171
[5] Georgi H and Glashow S L 1974 *Phys. Rev. Lett.* **32** 438
[6] Ma E 2013 *Phys. Rev. D* **88** 117702

GUT Scale Fermion Mass Ratios

Martin Spinrath
Institut für Theoretische Teilchenphysik, Karlsruhe Institute of Technology Engesserstraße 7, D-76131 Karlsruhe, Germany
E-mail: martin.spinrath@kit.edu

Abstract. We present a series of recent works related to group theoretical factors from GUT symmetry breaking which lead to predictions for the ratios of quark and lepton Yukawa couplings at the unification scale. New predictions for the GUT scale ratios y_μ/y_s, y_τ/y_b and y_t/y_b in particular are shown and compared to experimental data. For this comparison it is important to include possibly large supersymmetric threshold corrections. Due to this reason the structure of the fermion masses at the GUT scale depends on TeV scale physics and makes GUT scale physics testable at the LHC. We also discuss how this new predictions might lead to predictions for mixing angles by discussing the example of the recently measured last missing leptonic mixing angle θ_{13} making this new class of GUT models also testable in neutrino experiments.

1. Introduction

The Standard Model of particle physics (SM) has been tremendously successful in describing the available data about the fundamental building blocks of nature. To achieve this only nineteen parameters are needed but thirteen of them are related to the flavour sector. Adding neutrino masses to the SM introduces on top of that at least seven additional parameters. Such a large number of parameters seems not quite appropriate for the fundamental theory of nature. Furthermore, these parameters exhibit an unexpected pattern. The masses are strongly hierarchical and the mixing angles in the quark sector are small, whereas two of the three mixing angles in the lepton sector are large.

There are various attempts to reduce the number of parameters in the flavour sector. One very popular attempt is to introduce additional symmetries like flavour symmetries. We will focus here on models of supersymmetric Grand Unified Theories (SUSY GUTs), where some of the Yukawa couplings can be related to each other due to the underlying group structure. This is in fact not the first and biggest motivation for them. SUSY GUTs follow the old dream of a unified theory of everything. They contain candidates for dark matter and they can explain in a natural way the discretizing of electric charge and the light Higgs mass.

2. Fermion Mass Ratios

The main focus here lies on the flavour sector of SUSY GUTs. We will start our discussion with some theoretical considerations where we will discuss large low energy corrections to the fermion masses which have a significant impact on the fermion masses at the GUT scale. Then we will discuss how SUSY GUTs do predict certain ratios for the fermion mass ratios and present predictions from SU(5) and Pati-Salam. We will conclude this section with a brief description on how this might be related to the recently measured neutrino mixing angle θ_{13}.

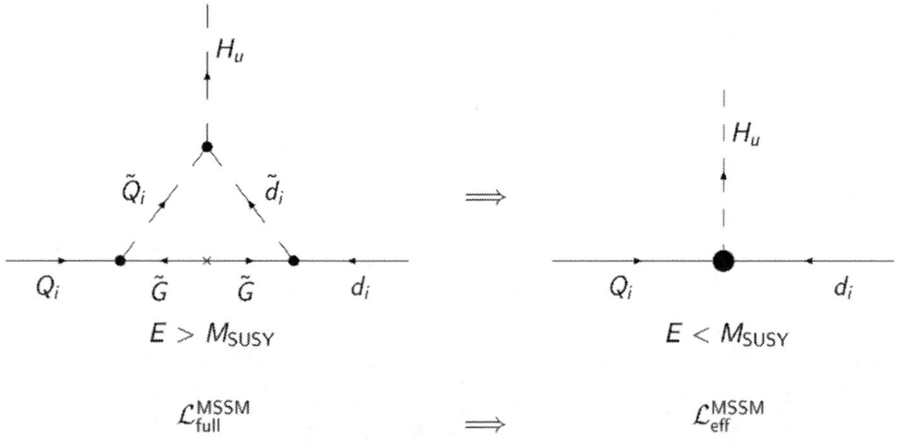

Figure 1. Matching the MSSM Yukawa couplings with the SM Yukawa couplings in a diagrammatic fashion.

2.1. Theoretical Considerations

Before we go into details we want to remark that we treat the running fermion masses as the product of the running Yukawa coupling times the low energy Higgs vev. Therefore there is a linear relation between fermion masses and Yukawa couplings and we will use the two terms interchangeably. Especially, if we take ratios of masses the Higgs vev drops out.

SUSY threshold corrections [1] can have a large impact on the high scale Yukawa couplings. After integrating out the SUSY particles at the SUSY scale M_{SUSY} one has to match the Yukawa couplings of the SM with the Yukawa couplings of the minimal supersymmetric standard model (MSSM). This is shown diagrammatically in Figure 1 and gives the relation

$$y_i^{\text{MSSM}} = \frac{y_i^{\text{SM}}}{\cos\beta(1+\epsilon_i \tan\beta)} , \tag{1}$$

where i stands for d, s, b, e, μ and τ and ϵ_i is a one loop factor. Explicit expressions for ϵ_i can be found in [2], where the importance of SUSY threshold corrections for the GUT scale Yukawa coupling ratios has been recently pointed out again. Since the one-loop factors ϵ_i appear with the possible large $\tan\beta$ the corrections can be quite significant, easily up to 30% and depending on the SUSY parameters they can appear with both signs.

We will present some results for the GUT scale fermion mass ratios in the next section. For the rest of this section we will discuss which predictions can be obtained within GUTs. On the renormalizable level in SU(5) one can get only $(Y_e)_{ij}/(Y_d)_{ij} = 1$ [3] or $(Y_e)_{ij}/(Y_d)_{ij} = -3$ [4], where $(Y_{d/e})_{ij}$ stands for the (i,j)-entry of the down-type quark/charged lepton Yukawa matrix. The same ratios with further relations to the up-quark sector are obtained in Pati-Salam (PS) models [5].

This options are very limited and in a Froggat-Nielsen approach [6] one would expect the smaller Yukawa couplings to be generated by higher-dimensional effective operators. Within this higher-dimensional operators one can add non-singlet fields which receive a higher vev. Below this symmetry breaking scale one could then end up with new predictions for Yukawa coupling ratios.

We studied this systematically for dimension five operators in SU(5) and PS [7, 8]. The considerations are based on the diagram in Figure 2. The fields A and C contain the MSSM matter fields and either B_1 or B_2 contains the MSSM Higgs fields. The other B field is a field

Journal of Physics: Conference Series **539** (2014) 012002 doi:10.1088/1742-6596/539/1/012002

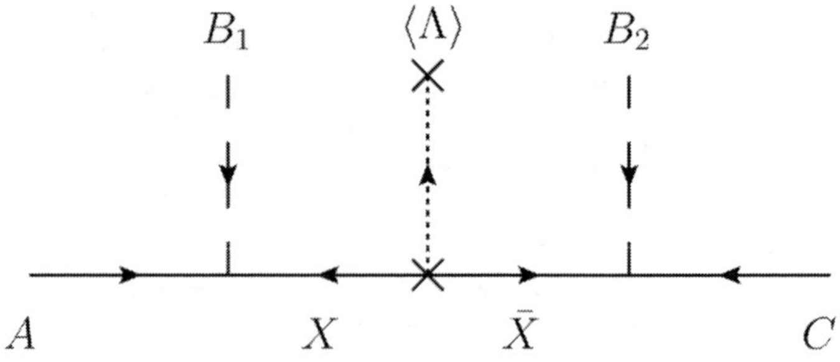

Figure 2. Diagrammatic representation of the operators giving Yukawa coupling ratios at the GUT scale. A and C are matter fields while B_1 and B_2 are Higgs fields. $\langle\Lambda\rangle$ represents the messenger mass term. If Λ is a total singlet one could directly write down a mass like in [7]. Otherwise Λ should transform as an adjoint of the GUT symmetry splitting the masses of the components of the messenger fields.

Table 1. Summary of possible SU(5) predictions for the GUT scale Yukawa coupling ratios $(Y_e)_{ij}/(Y_d)_{ij}$ taken from [8].

Operator Dimension	$(Y_e)_{ij}/(Y_d)_{ij}$
4	1
	-3
5	1/6
	-1/2
	-2/3
	1
	$\pm 3/2$
	-3
	9/2
	6
	-18

Table 2. Summary of possible PS predictions for the GUT scale Yukawa coupling ratios $((Y_e)_{ij}/(Y_d)_{ij}, (Y_u)_{ij}/(Y_d)_{ij}, (Y_\nu)_{ij}/(Y_u)_{ij})$ taken from [8].

Operator Dimension	$((Y_e)_{ij}/(Y_d)_{ij}, (Y_u)_{ij}/(Y_d)_{ij})$
4	(1, 1)
	(-3, 1)
5	(0, ± 1)
	(-1/3, ± 1)
	(1, ± 1)
	(3/2, ± 1)
	(-3, ± 1)
	(9, ± 1)

getting a heavy vev. In the first publication [7] we assumed the messenger fields X and $\bar{X}$ to have a direct mass term while in the second publication [8] we assumed the mass term to be generated dynamically, possibly from a non-singlet field offering even more options. The results are collected in Tables 1 and 2.

In the next section we will compare the theoretical calculations and possibilities with existing data.

2.2. Comparison to Data

In [7] we implemented a RGE running of the fermion masses to the GUT scale including the SUSY threshold corrections. For the SUSY breaking parameters we assumed three popular simplified scenarios, minimal anomaly mediated SUSY breaking (mAMSB), minimal gauge mediated SUSY breaking (mGMSB) and the constrained MSSM (CMSSM). Furthermore, we considered certain experimental constraints which were up to date at that time, for a detailed discussion of these constraints and the discussion of an additional dark matter constraint, see the original paper [7] or the author's PhD thesis [9].

The first thing to notice about our results summarized in Figure 3 is, that the common GUT assumptions of the GJ relation, $m_\mu/m_s = 3$ and complete third family Yukawa coupling unification, $m_t = m_b = m_\tau$, are somewhat challenged in our setup. Whereas the GJ relation is possible only for the mAMSB scenario for $\tan\beta \sim 30$, complete third family Yukawa coupling unification is incompatible with our results in all three scenarios. In the mAMSB case, where the corrections point in the right direction to achieve this relation without fine tuning, unification cannot be achieved because of non successful electroweak symmetry breaking and tachyons in the spectrum.

Nevertheless, we find valid alternatives to the standard assumptions. For the second generation we find the SU(5) relations $y_\mu/y_s = 9/2$ for $\tan\beta \sim 20$ and $y_\mu/y_s = 6$ for $\tan\beta \sim 50$ (not for mAMSB). The first of these is even valid for all three SUSY breaking scenarios (and smaller $\tan\beta$), which is simply due to the fact, that it emerges when the threshold corrections are small and therefore the dependence on the spectrum is weak.

For the third generation the SU(5) and PS relation $y_\tau/y_b = 1$ can be realized in mAMSB for $\tan\beta \sim 30$.[1] The SU(5) relation $y_\tau/y_b = 3/2$ appears in mGMSB for $\tan\beta \sim$ 30-45 and in CMSSM for $\tan\beta \sim$ 20-50. There are also alternatives in PS models, namely the relations $2y_t = 2y_b = y_\tau$ for $\tan\beta \sim 60$ in mGMSB and CMSSM and $y_t = 2y_b = 2y_\tau$ for $\tan\beta \sim 30$ in mAMSB. These relations are however generated from dimension six operators and have been obtained in [11], where a slightly different setup has been used.

Nevertheless, the discovery of a light Higgs boson [12] had a huge impact on the allowed MSSM parameter space. And the question for the naturalness of the Higgs mass in the MSSM became more pressing. In this context we studied the naturalness of the MSSM in a pMSSM like parametrization of the parameter space [13].

It turned out that the most effective way to reduce the amount of fine-tuning is to allow for non-universal gauging masses at the GUT scale. In Figure 4 we compare possible GUT scale ratios y_τ/y_b and y_t/y_b from a parameter scan before and after applying the Higgs mass constraint. The color coding gives the amount of fine-tuning. Without the Higgs mass constraint very low fine-tuning of order one is still allowed while after applying the constraint the fine-tuning is at least of order ten.

From a fine-tuning point of view one can also see that there is no clear preference for b-τ Yukawa coupling unification or $y_\tau/y_b = 3/2$. But it is very obvious how strong the Higgs mass constraint is. This shows again how important the LHC results are even for the possible physics at very large scales like the GUT scale.

[1] In our study we have focused on $\tan\beta$ between 20 and 60. We note that there can also be b-τ-unification of Yukawa couplings for $\tan\beta \sim 1$ (see e. g. [10]).

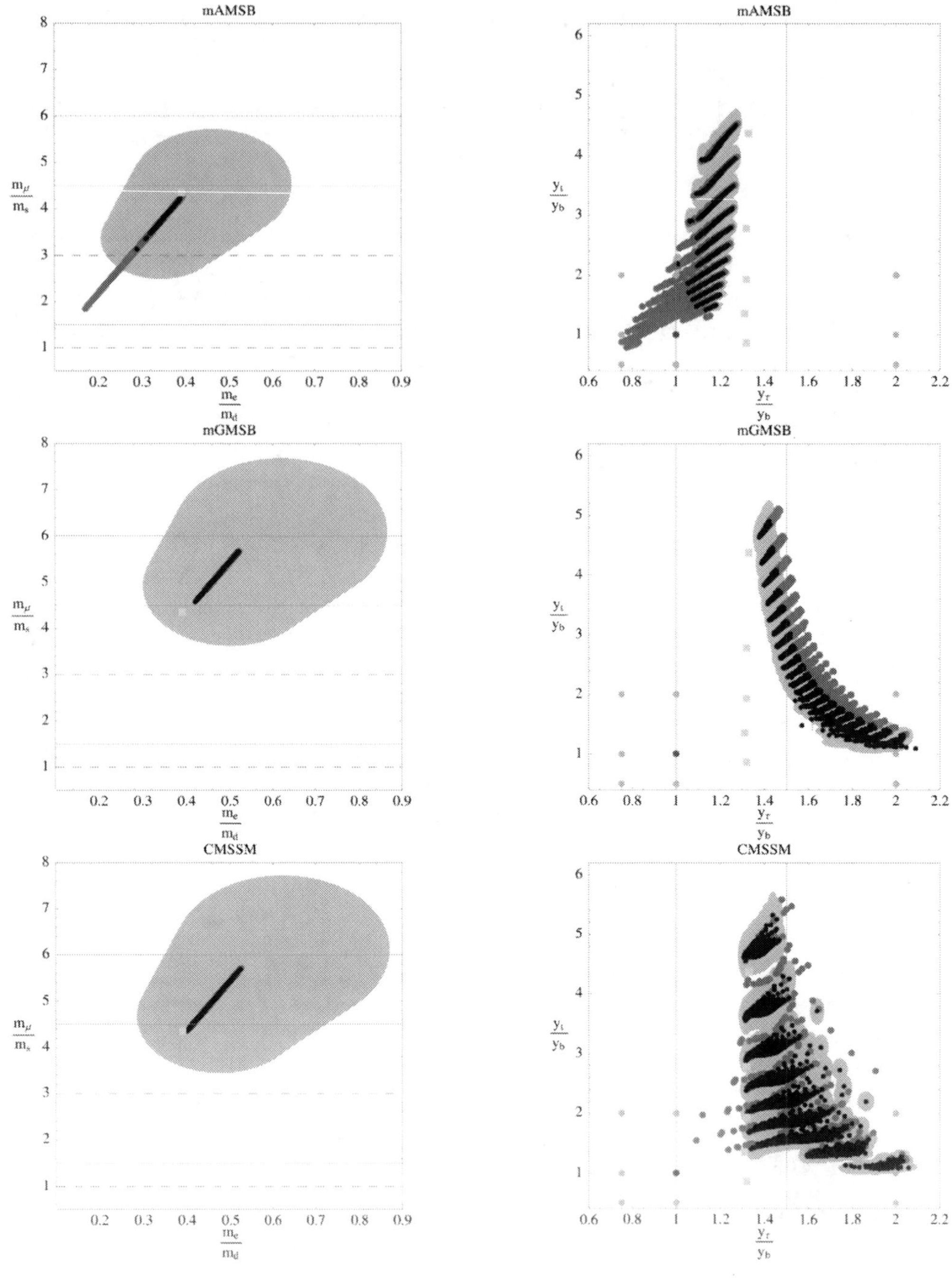

Figure 3. Results from a phenomenological scan for the mAMSB, mGMSB and CMSSM scenarios taken from [7]. The (red) black points are the (excluded) allowed points after applying experimental constraints. The grey regions indicate the experimental quark mass errors. The green lines are predictions from SU(5), the dashed lines from SU(5) and PS and the (light) blue points from PS (dimension-six operators). The yellow squads are the GUT scale Yukawa coupling ratios without including SUSY threshold corrections for $\tan\beta = 20$, 30, 40, 50 and 60 from top to bottom.

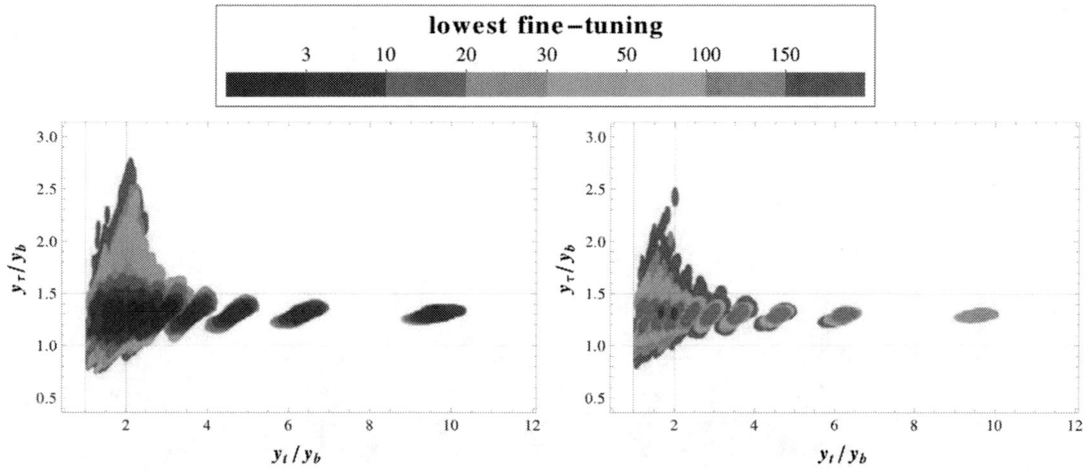

Figure 4. Lowest fine-tuning in the y_t/y_b-y_τ/y_b plane consistent with the experimental bounds as described in [13]. The Higgs mass constraint $m_h = 125.3 \pm 0.6$ GeV [12] and a theoretical uncertainty of ± 3 GeV [14] is (not) included in the (left) right plot. The 1σ errors on the quark masses [15] are taken into account by scaling the data points in the last plot correspondingly.

2.3. Connection to θ_{13}

In GUTs cannot only the fermion masses be related to each other but also relations between mixing angles are quite common. In this section we will discuss how such a connection can arise in SU(5) and its phenomenological implications based on [16].

In SU(5) the down-type quark and charged lepton matrix are identical up to transposition and order one Clebsch-Gordan (CG) coefficients, especially for the first two generations we find

$$\hat{\lambda}^D_{[12]} = \begin{pmatrix} a & b' \\ b & c \end{pmatrix} \text{ and } \hat{\lambda}^E_{[12]} = \begin{pmatrix} \alpha a & \beta b \\ \beta' b' & \gamma c \end{pmatrix} . \tag{2}$$

For the CG coefficients we can use the new relations from Table 1. Then one has to fit the four parameters a, b, b' and c to the masses of the down-type quarks and leptons of the first two generations. Furthermore, under the assumption that the up-quark mass matrix is very hierarchical the Cabibbo angle is dominated by the mixing in the down-quark sector so that we can also use this information.

Ususally, the system is then (over)determined for every choice of CG coefficients and we can predict the value of the 1-2 mixing angle in the lepton sector, θ^e_{12}. And now if the 1-3 mixing in the neutrino sector itself vanishes - which happens in certain well-motivated models - we can use the approximate relation for the physical θ_{13} mixing angle

$$\sin\theta_{13} = |U_{e3}| = \frac{\sin\theta^e_{12}\tan\theta_{23}}{\sqrt{1+\tan^2\theta_{23} - \sin^2\theta^e_{12}}} \approx \sin\theta^e_{12}\sin\theta_{23}, \tag{3}$$

which appears in the Pontecorvo-Maki-Nakagawa-Sakata (PMNS) matrix. In Figure 5 we show a histogram with the results of our scan over the possible combinations of CG coefficients. Indeed, we found several realistic combinations. And in [18] we have given an explicit SU(5)$\times T'$ flavour model with $\alpha = -3/2$, $\beta = 1$ and $\beta' = \gamma = 6$.

Finally, if the neutrino mixing matrix coincides with the bi-maximal (BM) [19] mixing matrix one has:

$$\sin^2\theta_{12} \simeq \frac{1}{2} + \cos\delta\,\sin\theta_{13} , \tag{4}$$

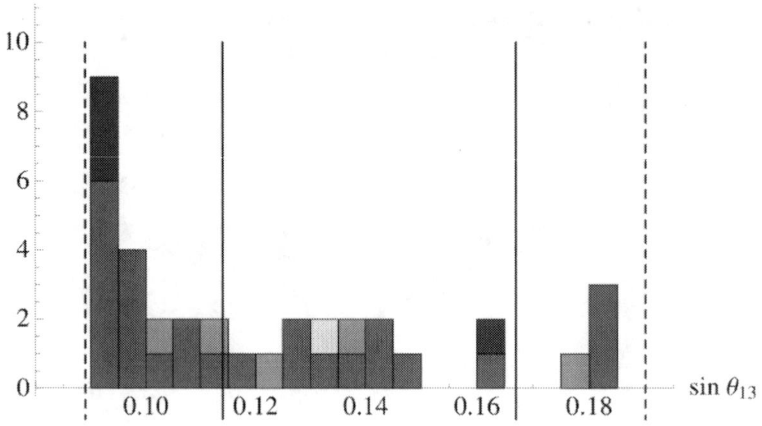

Figure 5. Graphical presentation of the results obtained. The vertical straight (dashed) lines denote the 1σ (2σ) allowed ranges of $\sin\theta_{13}$ taken from [17]. The colors correspond to certain simplifying assumptions, for more details, see [16].

and in the case of the tri-bimaximal mixing (TBM) form [20]:

$$\sin^2\theta_{12} \simeq \frac{1}{3} + \frac{2\sqrt{2}}{3}\cos\delta\,\sin\theta_{13}\;. \tag{5}$$

Since θ_{12} and θ_{13} are experimentally determined we get a constraint on the Dirac CP violating phase δ in the lepton sector.

We have displayed this constraint in Figure 6 in terms of $\cos\delta$ and the Jarlskog-Determinant J_{CP}. While the TBM case prefers large CP violation, $\cos\delta$ close to zero, the BM case prefers a small amount of CP violation, $\cos\delta \approx -1$. Therefore a precise measurement of δ can distinguish the two cases or exclude them altogether.

3. Summary and Conclusions

We have presented here a new class of GUT models where fermion mass ratios are predicted not from renormalizable, but non-renormalizable operators. This is natural because it can explain at the same time the smallness of most of the Yukawa couplings.

We have seen that these new predictions in very good agreement with current data and in fact a large portion of the MSSM parameter space points towards these new ratios instead of the conventional Yukawa unification or Georgi-Jarlskog relation.

One can read this ratios also from both directions. In a bottom-up approach it points towards the right fundamental theory which might involve the new fermion mass ratios. In a top-down approach the high-scale theory puts constraints on the low energy SUSY spectrum to be in agreement with the data.

Since we have entered now the LHC data and discovered a light Higgs boson the experimental constraints on the allowed parameter space become stronger and stronger so that we can hope to have soon a very concrete idea of the right fermion mass ratios at the GUT scale. Even more so, when we really discover SUSY at the LHC.

Beyond that we have also demonstrated how these ratios can have an impact on neutrino physics. While in SU(5) models employing the GJ relations θ_{13} often turns out to be to too small by a factor of three we can get the right θ_{13} by employing certain combinations of the new CG coefficients.

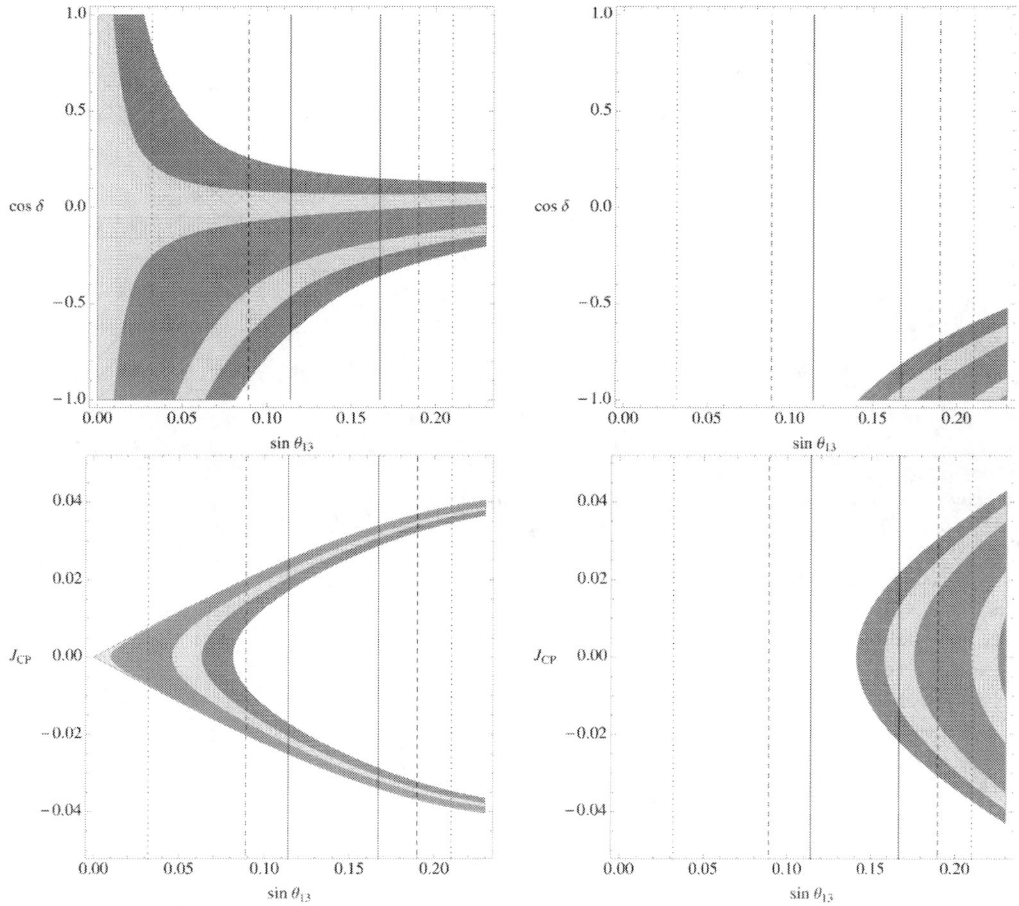

Figure 6. The cosine of the Dirac CP phase δ (upper panels) and the rephasing invariant J_{CP} (lower panels) as a function of $\sin\theta_{13}$ in the cases of tri-bimaximal (left panels) and bimaximal mixing (right panels) arising from the diagonalisation of the neutrino mass matrix taken from [16]. The green, yellow, orange regions correspond to the 1, 2, 3σ allowed ranges of $\sin^2\theta_{12}$. The vertical straight, dashed, dotted lines denote the 1, 2, 3σ allowed ranges of $\sin\theta_{13}$. The values of $\sin^2\theta_{12}$ and $\sin\theta_{13}$ are taken from [17].

Therefore we live in exciting times. Since the LHC started operating and since neutrino physics has entered the precision era we have made already huge progress in understanding the fundamental structure of nature. And in the near future we will learn even more from all the experimental efforts going on, like for instance the measurement of CP violation in the lepton sector.

References

[1] L. J. Hall, R. Rattazzi and U. Sarid, Phys. Rev. D **50** (1994) 7048 [arXiv:hep-ph/9306309]; M. S. Carena, M. Olechowski, S. Pokorski and C. E. M. Wagner, Nucl. Phys. B **426** (1994) 269 [arXiv:hep-ph/9402253]; R. Hempfling, Phys. Rev. D **49** (1994) 6168; T. Blazek, S. Raby and S. Pokorski, Phys. Rev. D **52** (1995) 4151 [arXiv:hep-ph/9504364].
[2] S. Antusch and M. Spinrath, Phys. Rev. D **78** (2008) 075020 [arXiv:0804.0717 [hep-ph]].
[3] H. Georgi and S. L. Glashow, Phys. Rev. Lett. **32** (1974) 438.
[4] H. Georgi and C. Jarlskog, Phys. Lett. B **86** (1979) 297.
[5] J. C. Pati and A. Salam, Phys. Rev. D **10** (1974) 275 [Erratum-ibid. D **11** (1975) 703].

Journal of Physics: Conference Series **539** (2014) 012002 doi:10.1088/1742-6596/539/1/012002

[6] C. D. Froggatt and H. B. Nielsen, Nucl. Phys. B **147** (1979) 277.
[7] S. Antusch and M. Spinrath, Phys. Rev. D **79** (2009) 095004 [arXiv:0902.4644 [hep-ph]].
[8] S. Antusch, S. F. King and M. Spinrath, arXiv:1311.0877 [hep-ph].
[9] M. Spinrath, arXiv:1009.2511 [hep-ph].
[10] G. Ross and M. Serna, Phys. Lett. B **664** (2008) 97 [arXiv:0704.1248 [hep-ph]].
[11] B. C. Allanach, S. F. King, G. K. Leontaris and S. Lola, Phys. Rev. D **56** (1997) 2632 [arXiv:hep-ph/9610517].
[12] G. Aad *et al.* [ATLAS Collaboration], Phys. Lett. B **716** (2012) 1 [arXiv:1207.7214 [hep-ex]]; S. Chatrchyan *et al.* [CMS Collaboration], Phys. Lett. B **716** (2012) 30 [arXiv:1207.7235 [hep-ex]].
[13] S. Antusch, L. Calibbi, V. Maurer, M. Monaco and M. Spinrath, JHEP **01** (2013) 187 [JHEP **1301** (2013) 187] [arXiv:1207.7236].
[14] A. Arbey, M. Battaglia, A. Djouadi and F. Mahmoudi, arXiv:1207.1348 [hep-ph].
[15] K. Nakamura *et al.* [Particle Data Group Collaboration], J. Phys. G **G37**, 075021 (2010).
[16] D. Marzocca, S. T. Petcov, A. Romanino and M. Spinrath, JHEP **1111** (2011) 009 [arXiv:1108.0614 [hep-ph]].
[17] G. L. Fogli, E. Lisi, A. Marrone, A. Palazzo, A. M. Rotunno, [arXiv:1106.6028 [hep-ph]].
[18] A. Meroni, S. T. Petcov and M. Spinrath, Phys. Rev. D **86** (2012) 113003 [arXiv:1205.5241 [hep-ph]].
[19] F. Vissani, hep-ph/9708483; V. D. Barger, S. Pakvasa, T. J. Weiler and K. Whisnant, Phys. Lett. B **437**, 107 (1998); A. J. Baltz, A. S. Goldhaber and M. Goldhaber, Phys. Rev. Lett. **81**, 5730 (1998); H. Georgi and S. L. Glashow, Phys. Rev. D **61**, 097301 (2000); I. Stancu and D. V. Ahluwalia, Phys. Lett. B **460**, 431 (1999).
[20] P. F. Harrison, D. H. Perkins and W. G. Scott, Phys. Lett. B **530**, 167 (2002); Phys. Lett. B **535**, 163 (2002); Z. Z. Xing, Phys. Lett. B **533**, 85 (2002); X. G. He and A. Zee, Phys. Lett. B **560**, 87 (2003); see also L. Wolfenstein, Phys. Rev. D **18**, 958 (1978).

Journal of Physics: Conference Series **539** (2014) 012003 doi:10.1088/1742-6596/539/1/012003

Hyperons in anisotropic neutron star

A. Sulaksono

Departemen Fisika, FMIPA, Universitas Indonesia, Depok 16424,Indonesia

E-mail: anto.sulaksono@sci.ui.ac.id

Abstract. The influence of modification of one of the assumptions in standard non-rotating static neutron star calculation i.e., the anisotropic pressure of neutron star matter with hyperons in its core within the framework of extended relativistic mean field model is investigated. It is found that the maximum mass and the redshift predictions are quite compatible with the recent observational constraints if this assumption is relaxed.

1. Introduction

The mass of pulsar J0348+0432, 2.01 $\pm$ 0.04 $M_\odot$ [1] puts a tight constraint on the equation of state (EOS) of dense matter in the core of neutron star. It means that the maximum mass M_{max} calculated using acceptable EOS should be greater than 2.0 $M_\odot$. This constraint is easy to fulfilled if NS core contain nucleons only. But if it is the case, the situation is physically not too interesting because the nuclear models that compatible with experimental data on hyper-nuclei predict the presence of hyperons at the density exceeding 2-3 ρ_0, where ρ_0 is baryon saturation density [2]. Unfortunately, the hyperonization of matter tends to soften the EOS of NS core due to the energetic nucleons are replaced by slowly moving hyperons. Brueckner-Hartree-Fock (BHF) models [3, 4] yield $M_{max} \sim$ 1.3-1.4 $M_\odot$ if hyperons are included in EOS and additional 3-body force also cannot help much to increase the predicted NS maximum mass [5]. In relativistic mean field (RMF) calculation using acceptable parameter sets, the situation is quite similar, $M_{max} >$ 2.0 $M_\odot$ can be reached only by tuning of the model parameters in hyperon sector [6, 7, 8, 9, 10]. If we anticipate however, that in the near future that heavier NS mass might be accurately measured such as black widow pulsar for example where the currently best estimate about 2.4 $M_\odot$ as well as other exotic such as boson condensates that plays a role also to soften the EOS, will be taken into account in the EOS of the NS core calculation, then the parameters fine tuning of RMF models with acceptable EOS as the way to solve this problem might exceed its limitation.

One of the standard assumptions in studying NS matter is the pressure of the matter is isotropic. However, there are arguments given (see Ref. [11] and references therein) that the pressure in the matter of NS can differ among the different directions (anisotropic) and this effect can be caused by many factors such as the presence of strong magnetic and electric fields, boson condensations, different kind of phase transition, the existence of solid core or super-fluidity and etc. It has been shown also that formally, the mixture of two fluids is mathematically equivalent to an anisotropic fluid [11]. Further, it has been shown in the past also that at high densities the nuclear matter pressure can be anisotropic (see for example [12]). Recent studies of the

Journal of Physics: Conference Series **539** (2014) 012003 doi:10.1088/1742-6596/539/1/012003

In this work, the effect of anisotropic pressure in NS mass-radius relation and the gravitational redshift has been investigated within extended version of relativistic mean field (ERMF) model [15] by taking into account the presence of hyperons in NS core.

2. Theoretical models

2.1. ERMF model

In this work we apply the NS crust EOS of Ref. [16] while the core EOS of this star is calculated by using ERMF model. ERMF model includes the contribution from the standard RMF nonlinear interaction for σ mason and the self-term ω as well as the addition cross interaction terms for σ, ω and ρ meson. The detailed and EOS derivation of this model are well documented in Refs. [17, 18]. In this model, baryons interact through the exchange of σ, ω, ρ and ϕ mesons, while the the baryons involved are nucleon (N) and hyperons (Λ, Σ and Ξ). Thus the total Lagrangian density, including lepton (l) to calculate the EOS of NS core can be written as

$$\mathcal{L} = \mathcal{L}_B^{\rm free} + \mathcal{L}_M^{\rm free} + \mathcal{L}_{BM}^{\rm lin} + \mathcal{L}^{\rm nonlin} + \mathcal{L}_l^{\rm free}. \tag{1}$$

The explicit expressions of each term of the Lagrangian density of Eq.(1) can be seen in Ref. [7]. The nucleons coupling constant and the nonlinear parameters (BSP parameter set) are taken from Ref. [15] while for determining the vector part of hyperons coupling constant $g_{\omega H}$ and $g_{\phi H}$, we consider standard prescription based on SU(6) symmetry [19]. For given values of $g_{\omega H}$ the scalar hyperons coupling strengths $g_{\sigma H}$ are usually obtained from potential depths for hyperons in the symmetric nuclear matter at the saturation density ρ_0 as,

$$U_H^{(N)}(\rho_0) = -g_{\sigma H}\sigma(\rho_0) + g_{\omega H}\omega(\rho_0), \tag{2}$$

where the values of potential depths $U_H^{(N)}$ chosen are as follows [19],

$$U_\Lambda^{(N)} = -28 \text{ MeV}, \quad U_\Sigma^{(N)} = +30 \text{ MeV} \quad \text{and} \quad U_\Xi^{(N)} = -18 \text{ MeV}. \tag{3}$$

The constituents composition in NS core is determined by using chemical potential balance, charge neutrality and conservation of baryon density conditions.

2.2. NS with anisotropic configuration

In this work, NS is assumed as a non-rotating spherical star with well known Schwarzchild like line element. The fluid anisotropy means that the radial pressure (p) differs from the transverse one (q) where the energy-momentum tensor takes following form:

$$T_{\mu\nu} = \epsilon u_\mu u_\nu + p k_\mu k_\nu + q[g_{\mu\nu} + u_\mu u_\nu - k_\mu k_\nu]. \tag{4}$$

Here $g_{\mu\nu}$ is the space-time metric, u_μ is the fluid 4-velocity, ϵ is energy density, k_μ is the unit radial vector where $u^\mu k_\mu$=0 and at the center of symmetry, the anisotropic pressure must vanish since k_μ is not defined there. Note that $g_{\mu\nu} + u_\mu u_\nu - k_\mu k_\nu$ is the projection tensor onto the 2-surfaces orthogonal to k_μ and u_μ. Inserted Eq.(4) into Einstein field equations and followed by manipulating all non-zero diagonal equations, we can obtain the Tolman-Oppenheimer-Volkoff (TOV) equations for anisotropic fluid as

$$\begin{aligned} \frac{dp}{dr} &= -G\frac{\epsilon M}{r^2}(1+\frac{p}{\epsilon})(1+\frac{4\pi r^3 p}{M})(1-\frac{2GM}{r})^{-1} - \frac{2(p-q)}{r}, \\ \frac{dM}{dr} &= 4\pi\epsilon r^2. \end{aligned} \tag{5}$$

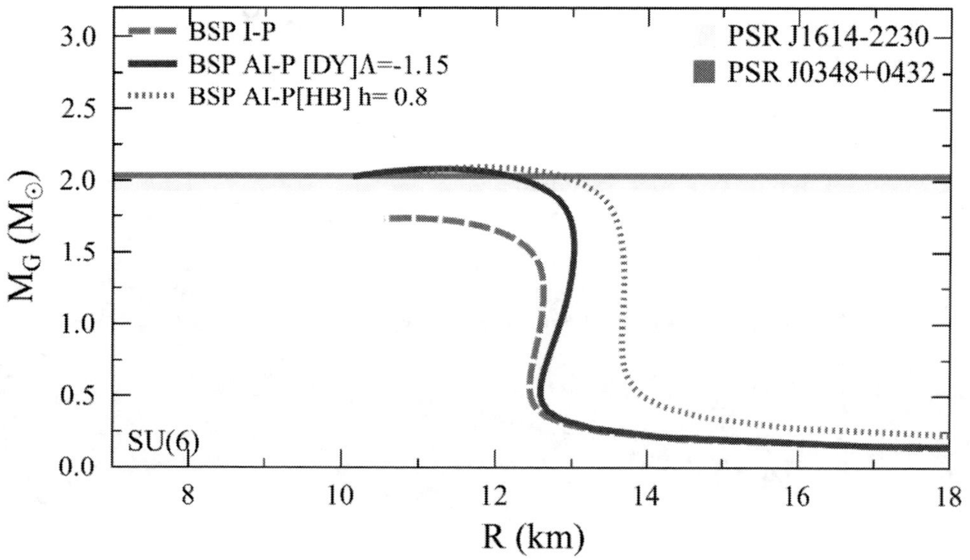

Figure 1. NS mass-radius relation. mass of PSR J0348+0432 is taken from Ref. [1] while PSR J1614-2230 form Ref. [20]. DY indicates the result obtained by using q of Ref. [14] and HB is the one obtained by using q of Ref. [13].

It is obvious that if we set p=q in first part of Eq. (5), it is back into standard TOV equation for isotropic case. In principle q should be determined by microscopic matter theory. Unfortunately, due to some difficulties, such as the difference of p and q is due to interplay many effects that mentioned in introduction, and the available microscopic models are developed in flat space-time and then its transferred form to curve space-time is not physically too satisfactory (see Refs. [13, 14], also references therein) etc, the q expression should be modeled. In this work, we investigate two models of q, one is introduced by the authors of Ref. [13] where it depends on $\frac{dp}{dr}$ and the other is from Ref. [14] where it depends directly on p where the explicit form of q :

$$\begin{aligned} q &= p[1-\Lambda(\frac{2MG}{r})], \quad \text{and} \\ q &= p+\frac{r}{2}\frac{(1-h)}{h}(\frac{dp}{dr}), \end{aligned} \tag{6}$$

where Λ and h are adjustable parameters. Note that general relativity predicts a redshift for photons leaving the surface of the star with strong gravitational field and the gravitational redshift of a non-rotating NS is given by $(1-\frac{2GM}{r})^{-1/2}-1$.

3. Results

The effects of the presence of hyperons and anisotropic pressure in NS mass-radius relation have been shown in Fig. 1. Note that here BSP parameter set of ERMF model [15] and standard SU(6) symmetry as well as the nuclear matter potential depths to determine hyperons coupling

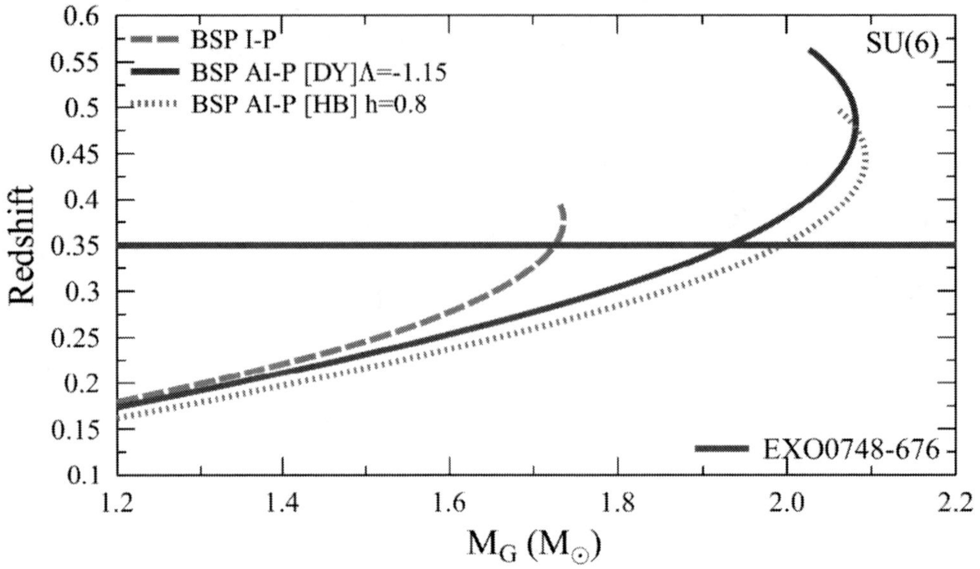

Figure 2. The gravitational redshift of NS as a function of mass.The horizontal line is redshift for low mass x-ray binary EXO0748-676 NS (Ref. [21]).

constant are used [19]. It can be observed that if we assumed the NS matter pressure is isotropic, the maximum mass prediction is unable to explain the mass of PSR J0348+0432 as well as PSR J1614-2230. However, If we relaxed the isotropic pressure assumption by allowing that the pressure of NS matter can differ among different direction, we can easily obtain maximum mass larger than the masses of PSR J0348+0432 and PSR J1614-2230 if $\Lambda \leq -1.15$ of DY [14] and $h \leq 0.8$ of HB [13] prescriptions are used in determining q. These results can be achieved without further adjusting the hyperons coupling constant, introducing exotic particle like weak interacting light boson (WILB) or modifying the nonlinear terms in strange meson sector [6, 7, 8, 9, 10]. It is interesting also to observe that the canonical NS (M=1.4 $M_\odot$) radius prediction among all results are significantly different. Therefore, the accurate measurements of NS radii can be used to discriminate the reliability of both prescriptions.

The redshift as function of mass for the isotropic and anisotropic pressures NS using HB and DY prescriptions with the presence of hyperons in the core is plotted in Fig. 2. The results are compared also to observational constraint from EXO0748-676 [21]. It can be seen that for all cases, the results are consistent with redshift =0.35 for mass greater than 1.7 $M_\odot$. These results are quite consistent with the suspected higher masses of accerting stars in x-ray binaries. It can be also observed that the effect of anisotropic pressure leads the redshift prediction at maximum mass is larger than that of isotropic one. The fact that the radius of maximum mass predicted HB is slightly larger than that of DY is the reason the redshift of the maximum mass of DY is larger than that of HB.

4. Conclusions

We have investigate the effect of anisotropic pressure on the mass-radius relation and the redshift of static NS with the presence of hyperons in NS core. The BSP parameter set of ERMF model [15] and standard SU(6) symmetry as well as the nuclear matter potential depths to determine hyperons coupling constant are used [19]. Without further adjusting the hyperons coupling constant, introducing exotic particle like weak interacting light boson (WILB) or modifying the nonlinear terms in strange meson sector [6, 7, 8, 9, 10], we can obtained easily maximum mass larger than the masses of PSR J0348+0432 and PSR J1614-2230 if $\Lambda \leq -1.15$ of DY [14] and $h \leq 0.8$ of HB [13] prescriptions are used in determining q. Furthermore, the effect of anisotropic pressure increases the redshift prediction at maximum mass.

References

[1] J. Antoniadis, *et al,* 2013 *Science* **340** 6131
[2] J. L. Zdunik, P. Haensel 2013 *Astron. Astrophys* **551** A61
[3] G. F. Burgio, H.-J. Schulze, A. Li 2011 *Phys. Rev. C* **83** 025804
[4] H.-J. Schulze, T. Rijken 2011 *Phys. Rev. C* **84** 035801
[5] I. Vidãna, D. Logoteta, C. Providência, A. Polls, I. Bombaci 2011 *Europhys.Lett* **94** 11002
[6] I. Bednarek, P. Haensel, J. L. Zdunik, M. Bejger, R. Mańka 2012 *Astron. Astrophys* **543** A157
[7] A. Sulaksono, B. K. Agrawal 2012 *Nucl. Phys. A* **895** 44
[8] W.-Z. Jiang, B.-A. Li, L.-W. Chen 2012 *Astrophys. J* **756** 56
[9] S. Weissenborn, D. Chatterjee, J. Schaffner-Bielich 2012 *Phys. Rev. C* **85** 065802
[10] A. Sulaksono, Marliana, Kasmudin 2011 *Mod. Phys. Lett. A* **26** 367
[11] L. Herrera, N. O. Santos 1997 *Phys. Rep.* **286** 53
[12] V. Canuto 1974 *Annu. Rev. Astron. Astrophys* **12** 167
[13] L. Herrera, W. Barreto 2013 *Phys. Rev. D* **88** 084022
[14] D. D. Doneva, S. S. Yazadjiev 2012 *Phys. Rev. D* **85** 124023
[15] B. K. Agrawal, A. Sulaksono, P.-G. Reinhard 2012 *Nucl. Phys. A* **882** 1
[16] T. Miyatsu, S. Yammamuro, K. Nakazaki 2013 *Astrophys. J* **777** 4
[17] R. Furnstahl, B. D. Serot, H-B. Tang 1996 *Nucl. Phys. A* **598** 539
[18] R. Furnstahl, B. D. Serot, H-B. Tang 1997 *Nucl. Phys. A* **615** 441
[19] J. Schaffner-Bielich, A. Gal 2000 *Phys. Rev. C* **62** 034311
[20] P. B. Demorest, *et al* 2010 *Nature* **467** 1081
[21] J. Cottam, F. Pearels, M. Mendez 2002 *Nature* **420** 51

Journal of Physics: Conference Series **539** (2014) 012004 doi:10.1088/1742-6596/539/1/012004

Exact Solution of Dirac Equation for q-Deformed Trigonometric Scarf potential with q-Deformed Trigonometric Tensor Coupling Potential for Spin and Pseudospin Symmetries Using Romanovski Polynomial

A. Suparmi[a,1], C. Cari[a], U.A. Deta[b] , A.S. Husein[a], and H. Yuliani[c,d]

[a]Physics Department of Graduate Program
Sebelas Maret University, Indonesia
Jl. Ir. Sutami 36A Kentingan, Surakarta 57126, Indonesia

[b]Department of Physics, Faculty of Mathematics and Natural Science
State University of Surabaya, Indonesia
Jl. Ketintang, Surabaya 60231, Indonesia

[c]Department of Physics Education, Faculty of Education and Teacher Training
State University of Palangka raya , Indonesia
Jl. Yos Sudarso, Palangka raya, Indonesia

[d]Prodi Tadris Fisika, Jurusan Tarbiah
STAIN Palangka raya , Indonesia
Jl. G. Obos, Palangka raya, Indonesia

[1]E-mail: suparmiuns@gmail.com

Abstract. The bound state solutions of Dirac equation for q-deformed trigonometric Scarf potential with q-deformed trigonometric cotangent and cosecant tensor coupling potential under spin and pseudospin symmetric limits are investigated using Supersymmetric Quantum Mechanics (SUSY QM) method. The new tensor potentials proposed is inspired by superpotential form in SUSY quantum mechanics. The Dirac equations for Scarf potential coupled by new tensor potential in the pseudospin and spin symmetric cases reduce to Schrodinger type equations for shape invariant potential since the proposed new potentials are equivalent to the superpotential of q-deformed trigonometric Scarf potential. The relativistic wave functions are exactly obtained by using SUSY operator method and the relativistic energy equation are exactly obtained by using SUSY method and the idea of shape invariance in the approximation scheme of centrifugal term. The new tensor potential causes the energy degeneracies is omit both for pseudospin and spin symmetric case.

Journal of Physics: Conference Series **539** (2014) 012004 doi:10.1088/1742-6596/539/1/012004

1. Introduction

The exact solutions of Dirac equations play important roles in relativistic quantum mechanics since they provide all important information of the system under consideration. Dirac equation for central potentials coupled with Yukawa-type and Coulomb-type tensor potentials and its modification have been solved exactly and applied in quantum chemistry and high energy physics. To describe the motion of spin half particles, some authors have applied various solution methods such as Nikiforov-Uvarof (NU) method [1-5], factorization methods and SUSY QM [6], hypergeometric and confluent hypergeometric method [7-8], and asymptotic iteration method [9]. Dirac equation for central potential such as Hulthen potential, Morse potential, Yukawa potential, attractive exponential potential, Poschl-Teller potential, etc. together with new tensor coupling potential have exact solution only for spin orbit quantum number, κ is zero, but for non zero κ the approximation scheme of centrifugal term has to be taken account to give exact solution [1, 2, 10-15]. The approximation scheme of the centrifugal term was proposed by Greene and Aldrich [16] and this approximation works well for trigonometric, hyperbolic and exponential potentials. From the observation, the expression of the tensor coupling potentials under the approximation scheme for centrifugal term are similar to the expression of the corresponding potential.

The new tensor coupling potentials proposed is inspired by the SUSY algebraic structure for Dirac equation under spin and pseudospin limit for certain potentials such as trigonometric potential, in particular. For these potentials, it is proposed that the tensor coupling potentials are its superpotential. However, these new tensor coupling potentials are so specific therefore can not be widely used as Coulomb-type and Yukawa-type tensor potentials. But it is worthy to be explored to provide new decription of the motion of spin-half particles.

The Dirac equations for some potentials have been solved in the cases of spin symmetry and pseudospin symmetry [5, 8, 17-25]. The spin symmetry occurs when the different between repulsive vector potential with the attractive scalar potential is equal to constant, while the pseudospin symmetry arises when the sum of the scalar potential with vector potential is equal to constant. Spin symmetric and pseudospin symmetric concepts have been used to study the aspect of deformed and superdeformation nuclei in nuclear physics. The concept of spin symmetry has been applied to the spectrum of meson and antinucleon [26], and the pseudospin symmetric concept is used to explain the quasi degeneracy of the nucleon doublets [27], exotic nuclei [28], super-deformation in nulei [29], and to establish an affective nuclear shell-model scheme [30].

In this paper, we propose new tensor coupling potential as a function of trigonometric and hyperbolic terms expressed as

$$U(r) = -a\left(V_2 \cot ar + V_3 \csc ar\right) \tag{1}$$

The negative q-deformed trigonometric cotangent plus cosecant tensor potential is a little more negative than Coulomb-like tensor potential. Screened Coulomb potential was originally used to model strong nuleon-nucleon interactions caused by the exchange in nuclear physics[31-33]. The Dirac equation with this new tensor coupling potential can be solved exactly under approxomation scheme of centrifugal term only for q-deformed trigonometric Scarf potential since the combination of tensor potential together with centrifugal term and q-deformed trigonometric Scarf potential resulting shape invariant potential in the Dirac equation. The relativistic energy and wave functions of this new tensor potential together with Scarf potential plus centrifugal term are analyzed using Romanovski polynomials. The trigonometric Scarf potential is potential model used to explain strong and electromagnetic interactions.

Finite Romanovski polynomial is traditional method which consist of reducing Schrodinger equation by an appropriate variable substitution to a form of generalized hypergeometric equation [34]. The polynomial was discovered by Sir E. J. Routh [35] and rediscovered 45 years later by V. I. Romanovski[36]. The notion "finite" refers to the observation that, for any given set of parameters (i.e. in any potential) only a finite of polynomials appear orthogonal [37-39]. From the observation only few researcher used Romanovski polynomials to analyze energy spectra and wave function for certain potentials [37, 39, 40-42].

This paper is organized as follows. Basic theory of Dirac equation is presented in section 2, and the Romanovski polynomial method as analysis method is presented in section 3. Section 4 presents the research results and discussion, and finally brief conclusion and acknowledgement are presented in section 5.

2. Basic Theory

2.1. Basic Equations of Dirac Spinors

The motion of nucleon with mass M in a repulsive vector potential $V_V(\vec{r})$ and an attractive scalar potential $V_S(\vec{r})$ and also coupled by a tensor potential $U(r)$ is described by Dirac equation given as [1, 2, 5, 8, 20, 22, 32]

$$\{\vec{\alpha}.\vec{p}+\beta(M+V_S(\vec{r}))-i\beta\vec{\alpha}.\vec{r}U(r)\}\psi(\vec{r})=\{E-V_V(\vec{r})\}\psi(\vec{r}) \quad (2)$$

where E is the relativistic energy and $\vec{p}$ is the three dimensional momentum operator, $-i\nabla$,

$$\vec{\alpha}=\begin{pmatrix}0 & \sigma\\ \sigma & 0\end{pmatrix} \quad (3)$$

and

$$\beta=\begin{pmatrix}I & 0\\ 0 & -I\end{pmatrix} \quad (4)$$

with σ is three dimensional Pauli matrices, I is 2×2 identity matrix. Here we consider the matrix potential in equation (1) as spherically symmetric potential, they do not only depend on the radial coordinate $r=|\vec{r}|$ and we have taken $\hbar=1$, c = 1. The Dirac equation expressed in equation (3) is invariant under spatial inversion, and therefore its eigenstates have definite parity. By writing the spinor as

$$\psi(\vec{r})=\begin{pmatrix}\zeta(\vec{r})\\ \varphi(\vec{r})\end{pmatrix}=\begin{pmatrix}\dfrac{F_{nK}(r)}{r}Y^{l}_{jm}(\theta,\phi)\\ i\dfrac{G_{nK}(r)}{r}Y^{\tilde{l}}_{jm}(\theta,\phi)\end{pmatrix} \quad (5)$$

where $\zeta(\vec{r})$ is Dirac spinor of upper (large) component and $\varphi(\vec{r})$ is Dirac spinor of lower (small) component, $Y^{l}_{jm}(\theta,\phi)$ is spin spherical harmonics, $Y^{\tilde{l}}_{jm}(\theta,\phi)$ is pseudospin spherical harmonics, l is orbital quantum number, $\tilde{l}$ pseudo orbital quantum number, and m is the projection of the angular momentum on the z-axis. The Dirac Hamiltonian in a spherical field commutes with total angular momentum operator $\vec{J}$ and spin orbit coupling operator $\vec{K}$ with its spin orbit quantum number $K=-\beta(\vec{\sigma}.\vec{L}+1)$, $\vec{L}$ is the usual orbital angular momentum. The eigenvalues of the spin orbit coupling operator are $\kappa=(J+1/2)>0$ for unaligned spin $(p_{1/2},d_{3/2},...)$ and

$\kappa = -(J+1/2) < 0$ for aligned spin $(s_{1/2}, p_{3/2}, ...)$. Therefore the conservative quantities are consisting of set of $\vec{H}, \vec{K}, \vec{J}, J_z$.

By inserting equations (5), and (4), into equation (1) we have

$$\begin{pmatrix} 0 & \sigma \\ \sigma & 0 \end{pmatrix} p \begin{pmatrix} \zeta(r) \\ \varphi(r) \end{pmatrix} + \begin{pmatrix} I & 0 \\ 0 & -I \end{pmatrix} (M + V_S(\vec{r})) \begin{pmatrix} \zeta(r) \\ \varphi(r) \end{pmatrix} - i\beta\vec{\alpha}.\hat{r}U(r) \begin{pmatrix} \zeta(r) \\ \varphi(r) \end{pmatrix} = (E - V_V(\vec{r})) \begin{pmatrix} \zeta(r) \\ \varphi(r) \end{pmatrix} \quad (6)$$

From equation (6) we obtain the coupled first order differential equation given as

$$\left(\frac{d}{dr} + \frac{\kappa}{r} - U(r)\right) F_{n\kappa}(r) = \left(M + E_{n\kappa} - V_V(r) + V_S(r)\right) G_{n\kappa}(r) \quad (7)$$

and

$$\left(\frac{d}{dr} - \frac{\kappa}{r} + U(r)\right) G_{n\kappa}(r) = \left(M - E_{n\kappa} + V_V(r) + V_S(r)\right) F_{n\kappa}(r) \quad (8)$$

where $F_{n\kappa}(r)$ is the upper component of Dirac spinor and $G_{n\kappa}(r)$ is the lower component of Dirac spinor. From equations (7) and (8) we get the upper and lower radial part of the Dirac equation,

$$\left\{\frac{d^2}{dr^2} - \frac{\kappa(\kappa+1)}{r^2} + \frac{2\kappa}{r}U(r) - U^2(r) - \frac{dU}{dr} + \frac{d\Delta}{dr}\left(\frac{d}{dr} + \frac{\kappa}{r} - U(r)\right) / \left(M + E_{n\kappa} - \Delta(r)\right)\right\} F_{n\kappa}(r) + \quad (9)$$
$$\left(M + E_{n\kappa} - \Delta(r)\right)\left(E_{n\kappa} - M - \Sigma(r)\right) F_{n\kappa}(r) = 0$$

$$\left\{\frac{d^2}{dr^2} - \frac{\kappa(\kappa-1)}{r^2} + \frac{2\kappa}{r}U(r) - U^2(r) + \frac{dU}{dr} + \frac{d\Sigma}{dr}\left(\frac{d}{dr} - \frac{\kappa}{r} + U(r)\right) / \left(M - E_{n\kappa} + \Sigma(r)\right)\right\} G_{n\kappa}(r) \quad (10)$$
$$+\left(M + E_{n\kappa} - \Delta(r)\right)\left(E_{n\kappa} - M - \Sigma(r)\right) G_{n\kappa}(r) = 0$$

with $\kappa(\kappa+1) = l(l+1)$ for upper spinor component, $\kappa(\kappa-1) = \bar{l}(\bar{l}+1)$ for lower spinor component, $\bar{l}$ is the pseudo orbital quantum number, $\Sigma(r) = V_V(r) + V_S(r)$ is the sum of scalar and vector potentials, and $\Delta(r) = V_V(r) - V_S(r)$ is the different between vector potential and scalar potential. There are two special cases for Dirac equation, pseudospin symmetric case and spin symmetric case.

Pseudospin symmetry occurs when $\Sigma(r) = V_V(r) + V_S(r) = C_{ps}$ with C_{ps} is constant, therefore $\frac{d\Sigma}{dr} = 0$ and the different between vector and scalar potentials $\Delta(r)$ is set to be equal with the given potential,

$$\Delta(r) = V_V(r) - V_S(r) = V(r) \quad (11)$$

In the limit of pseudospin symmetry we have the Dirac equation for lower component of Dirac spinor obtained from equation (8) as

$$\left\{\frac{d^2}{dr^2} - \frac{\kappa(\kappa-1)}{r^2} + \frac{2\kappa}{r}U(r) - U^2(r) + \frac{dU}{dr} + V(r)\left(M - E_{n\kappa} + C_{ps}\right)\right\} G_{n\kappa}(r) = \left(M + E_{n\kappa}\right)\left(M - E_{n\kappa} + C_{ps}\right) G_{n\kappa}(r) \quad (12)$$

From $\kappa(\kappa-1) = \bar{l}(\bar{l}+1)$ one gets the values of κ as $\kappa = -\bar{l} = -(j+1/2)$ for $\kappa < 0$ which is associated with aligned spin and $\kappa = (\bar{l}+1) = j+1/2$, $\kappa > 0$ for unaligned spin. In general the pseudo orbital quantum number is written as $\bar{l} = l - \kappa/|\kappa|$. These conditions imply that the total angular momentum $j = \bar{l} \pm \frac{1}{2}$ that causes the state to be degenerated for $\bar{l} \neq 0$. The Dirac equation expressed in equation (12) is Schrodinger-like equation and therefore can be solved exactly only if the effective potential,

$$V_{ef}=\frac{\kappa(\kappa-1)}{r^2}-\frac{2\kappa}{r}U(r)+U^2(r)-\frac{dU}{dr}-V(r)\left(M-E_{n\kappa}+C_{ps}\right)=\varphi^2(r)-\varphi'(r)-V(r)\left(M-E_{n\kappa}+C_{ps}\right) \quad (13)$$

is shape invariant, with $\varphi(r)=U(r)-\frac{\kappa}{r}$. Equation (13) shows that the effective potential is mixture of two potentials, $V_L=\varphi^2(r)-\varphi'(r)$, and $V(r)\left(M-E_{n\kappa}+C_{ps}\right)$ with $V(r)$ is given potential. Therefore equation (13) will be shape invariant potential if in the approximation scheme of the centrifugal term, $\varphi(r)$ has the form of the superpotential of the potential system $V(r)$. So for the tensor potential expressed in equations (1), the suitable given potentials in Dirac equation expressed in equation (12) are the trigonometric.

On the other hand, the spin symmetry occurs when the different between vector and scalar potentials is constant, $\Delta(r)=V_V(r)-V_S(r)=C_s$, and the sum of vector and scalar potentials is expressed as $\Sigma(r)=V_V(r)+V_S(r)=V(r)$, therefore in the spin symmetric limit we get the upper component of Dirac spinor obtained from equation (9) given as

$$\left\{\frac{d^2}{dr^2}-\frac{\kappa(\kappa+1)}{r^2}+\frac{2\kappa}{r}U(r)-U^2(r)-\frac{dU}{dr}-V(r)\left(M+E_{n\kappa}-C_s\right)\right\}F_{n\kappa}(r)=\left(M+E_{n\kappa}-C_s\right)\left(M-E_{n\kappa}\right)F_{n\kappa}(r) \quad (14)$$

with $\kappa(\kappa+1)=l(l+1)$ that leads to $\kappa=l=j+\frac{1}{2}$, $\kappa>0$ for unaligned spin and $\kappa=-(l+1)=-(j+\frac{1}{2})$, $\kappa<0$ for aligned spin. The Dirac equation expressed in equation (14) is Schrodinger-like equation and therefore can be solved exactly only if the effective potential, Vef,

$$V_{ef}=\frac{\kappa(\kappa+1)}{r^2}-\frac{2\kappa}{r}U(r)+U^2(r)+\frac{dU}{dr}+V(r)\left(E_{n\kappa}+M-C_s\right)=\varphi^2(r)+\varphi'(r)+V(r)\left(E_{n\kappa}+M-C_s\right) \quad (15)$$

is shape invariant, with $\varphi(r)=U(r)-\frac{\kappa}{r}$. As in equation (13), equation (15) is the effective potential of equation (14) which is mixture of two potentials, first part is centrifugal term plus tensor potential, $V_L=\varphi^2(r)+\varphi'(r)$, and the second term, $V(r)$, is the sum of the attractive and repulsive potential. In the approximation scheme of the centrifugal term the effective potential Vef is shape invariance if $\varphi(r)$ has the form of the superpotential of the potential system $V(r)$.

The energy spectra and the upper component wavefunction of Dirac spinor for the spin symmetric case is obtainable from the energy spectra and the lower component wave function of Dirac spinor for pseudospin symmetric case [39]. By comparing the Dirac equation for pseudospin symmetry in equation (12) and spin symmetry in equation (14) we obtain mapping parameters produced from energy equations for pseudospin and spin symmetries as follows

$$G_{n\kappa}(r)\rightarrow F_{n\kappa}(r),\ \Delta(r)\rightarrow-\Sigma(r)\text{ or } V_0\rightarrow -V_0;\ V_1\rightarrow -V_1,\ E_{n\kappa ps}\rightarrow -E_{n\kappa s},\ C_{ps}\rightarrow -C_s \text{ and } \kappa\rightarrow\kappa+1 \quad (16)$$

Both Dirac equations for pseudospin symmetry and spin symmetry in equations (12) and (14) are solved using SUSY QM method. In addition, by using equation (16) we can also map the energy equation from pseudospin symmetric case into spin symmetric case.

2.2. *Review of q-deformed trigonometric function*

For spin and pseudospin symmetry the Dirac equation for q-deformed trigonometric Scarf potential within the q-deformed trigonometric cotangent plus cosecant type tensor reduces to Schrodinger-type equation therefore it can be solved using Romanovski polynomials. Cotangent plus cosecant tensor potential behaves like Coulomb-like tensor potential therefore it can be used to describe the nucleon-nucleon interaction [31-33]. The q-deformed trigonometric Scarf potential and the q-deformed trigonometric cotangent plus cosecant type tensor given as

$$V(r)=a^2\left\{\frac{V_0}{\sin_q^2 ar}-\frac{V_1\cos_q ar}{\sin_q^2 ar}\right\} \tag{17}$$

$$U(r)=-a\left(V_2\cot_q ar+V_3\csc_q ar\right) \tag{18}$$

V_0 and V_1 describe the depth of the trigonometric function well potential and are positives, $V_1 > V_0$, a is a positive parameter which to control the width or the range of the potential well, q is the deformation of the potential, $q > 0$, V_2 and V_3 are the strength of the nucleon forces, a is the range of nucleon force, M is the mass of the particle, and $0<r<\infty$.

The q-deformed trigonometric function is formulated in the same way with the formulation of q-deformed hyperbolic function introduced by Arai [44] some years ago, in accordingly the q-deformed trigonometric function is defined as the definition of trigonometric function as follows:

$$\sin_q ar=\frac{e^{iar}-qe^{-iar}}{2};\ \cos_q ar=\frac{e^{iar}+qe^{-iar}}{2};\ \sin_q^2 ar+\cos_q^2 ar=q \tag{19}$$

$$\tan_q ar=\frac{\sin_q ar}{\cos_q ar};\ \sec_q ar=\frac{1}{\cos_q ar};\ 1+\tan_q^2 ar=q\sec_q^2 ar \tag{20}$$

$$\frac{d\sin_q ar}{dr}=a\cos_q ar;\ \frac{d\tan_q ar}{dr}=qa\sec_q^2 ar \tag{21}$$

By a convenient translation of the spatial variable, one can transform the deformed potentials into the corresponding non-deformed ones or vice-versa. In analogy to the translation of spatial variable for hyperbolic function introduced by Dutra [45] we propose the translation of spatial variable for trigonometric function as follows

$$r\to r+\frac{\ln\sqrt{q}}{i\alpha},\text{ and }\ r\to r-\frac{\ln\sqrt{q}}{i\alpha} \tag{22}$$

and then by inserting equation (22) into equations (19) and (20) we have

$$\sin_q \alpha r\to\sqrt{q}\sin\alpha r;\ \ \cos_q \alpha r\to\sqrt{q}\cos\alpha r;\ \text{or}\ \sin\alpha r\to\frac{\sin_q \alpha r}{\sqrt{q}};\ \ \cos\alpha r\to\frac{\cos_q \alpha r}{\sqrt{q}}; \tag{23}$$

The translation of spatial variable in equation (22) can be used to map the energy and wave function of non-deformed potential toward deformed potential of Scarf potential [46].

3. Method of Analysis

The method used to solve the Dirac equation in the limit of spin symmetric and pseudospin symmetric cases is the Romanovski polynomials since the Dirac equations for limited condition, when spin and pseudospin symmetry arise, reduce to one dimensional Schrodinger-like equation. The one dimensional second order differential equation satisfied by Romanovski polynomials is developed based on hypergeometric differential equation. One dimensional Schrodinger equation of potential of interest reduces to the differential equation of Romanovski polynomial by appropriate variable and wave function substitutions. The one dimensional Schrodinger equation is given as

$$-\frac{\hbar^2}{2M}\frac{\partial^2\Psi(x)}{\partial x^2}+V(x)\Psi(x)=E\Psi(x) \tag{24}$$

where $V(x)$ is an effective potential which is mostly shape invariant potential. By suitable variable substitution $x = f(s)$ equation (24) changes into generalized hypergeometric type equation expressed as

$$\frac{\partial^2\Psi(s)}{\partial s^2}+\frac{\tilde{\tau}(s)}{\sigma(s)}\frac{\partial\Psi(s)}{\partial s}+\frac{\tilde{\sigma}(s)}{\sigma^2(s)}\Psi(s)=0 \tag{25}$$

with $\sigma(s)$ and $\tilde{\sigma}(s)$ are mostly polynomials of order two, $\tilde{\tau}(s)$ is polynomial of order one, s, $\sigma(s)$, $\tilde{\sigma}(s)$, and $\tilde{\tau}(s)$ can have any real or complex values [47]. Equation (25) is solved by variable separation method. By introducing new wave function in equation (25),

$$\Psi_n(r)=g_n(s)=(1+s^2)^{\frac{\beta}{2}}e^{\frac{-\alpha}{2}\tan^{-1}s}D_n^{(\beta,\alpha)}(s) \tag{26}$$

We obtain a hypergeometric type differential equation, which can be solved using finite Romanovski polynomials [37, 38, 39, 47] is expressed as

$$\sigma(s)y''(s)+\tau(s)y'(s)+\lambda y(s)=0 \tag{27}$$

with $\sigma(s)=as^2+bs+c;\ \tau=fs+h$ and $-\{n(n-1)+2n(1-p)\}=\lambda=\lambda_n$ (28)

and $y_n=R_n^{(p,q')}(s)=D_n^{(\beta,\alpha)}(s)$ (29)

For Romanovski polynomials, the values of parameters in equation (28) are
$a = 1, b = 0, c = 1, f = 2(1-p)$ and $h = q'$ with $p > 0$ (30)
therefore equation (27) is rewritten as

$$(1+s^2)\frac{\partial^2 R_n^{(p,q')}}{\partial s^2}+\{2s(-p+1)+q'\}\frac{\partial R_n^{(p,q')}(s)}{\partial s}-\{n(n-1)+2n(1-p)\}R_n^{(p,q')}(s)=0 \tag{31}$$

Equation (31) which is obtained from equation (27) by applying the specific condition for Romanovski polynomials expressed in equation (30) is second order differential equation satisfied by Romanovski polynomials. Equation (27) is described in the textbook by Nikiforov-Uvarov [M] where it is cast into self adjoint form and its weight function, w (s), satisfies Pearson differential equation

$$\frac{d(\sigma(s)w(s))}{ds}=\tau(s)w(s) \tag{32}$$

The weight function, $w(s)$, is obtained by solving the Pearson differential equation expressed in equation (32) and by applying condition in equations (28) and (30), given as

$$w^{(p,q')}(s)=(1+s^2)^{-p}e^{q'\tan^{-1}(s)} \tag{33}$$

The corresponding polynomials are classified according to the weight function, and are built up from the Rodrigues representation which is presented as

$$y_n=\frac{B_n}{w(s)}\frac{d^n}{ds^n}\left\{\left(as^2+bs+c\right)^n w(s)\right\} \tag{34}$$

with Bn is a normalization constant, and for $\sigma(s)>0$ and $w(s) > 0$, $y_n(s)$'s are normalized polynomials and are orthogonal with respect to the weight function $w(s)$ within a given interval (s_1, s_2), which is expressed as

$$\int_{-\infty}^{\infty} w(s)y_n(s)y_{n'}(s)ds=\delta_{nn'} \tag{35}$$

This weight function in equation (33) first reported by Routh [35] and then by Romanovski [36]. The polynomial associated with equation (31) are named after Romanovski and will be denoted by $R_n^{(p,q)}(s)$. Due to the decrease of the weight function by s^{-2p}, integral of the type

$$\int_{-\infty}^{\infty} w^{(p,q')} R_n^{(p,q')}(s) R_{n'}^{(p,q')}(s) ds \tag{36}$$

will be convergent only if $n' + n < 2p - 1$ (37)
This means that only a finite number of Romanovski polynomials are orthogonal, and the orthogonality integral of the polynomial is expressed similar to the equation (35) where $y_n = R_n^{(p,q')}(s)$.

The Romanovski polynomials obtained from Rodrigues formula expressed in equation (34) with the corresponding weight function in equation (33) is expressed as

$$R_n^{(p,q')}(s) = D_n^{(\beta,\alpha)}(s) = \frac{1}{(1+s^2)^{-p} e^{q'\tan^{-1}(s)}} \frac{d^n}{ds^n}\left\{(1+s^2)^n (1+s^2)^{-p} e^{q'\tan^{-1}(s)}\right\} \tag{38}$$

If the wave function of the nth level in equation (26) is rewritten as

$$\Psi_n(r) = \frac{1}{\sqrt{\frac{df(s)}{ds}}} (1+s^2)^{\frac{-p}{2}} e^{\frac{q'}{2}\tan^{-1}(s)} R_n^{(p,q')}(s) \tag{39}$$

then the orthogonality integral of the wave functions expressed in equation (39) gives rise to orthogonality integral of the finite Romanovski polynomials, that is given as

$$\int_0^{\infty} \Psi_n(r) \Psi_{n'}(r) dr = \int_{-\infty}^{\infty} w^{(p,q')} R_n^{(p,q')}(s) R_{n'}^{(p,q')}(s) ds \tag{40}$$

In this case the values of p and q' are not n-dependence where n is the degree of polynomials. However, if either equation (35) or (37) is not fulfilled then the Romanovski polynomials is infinity [37].

4. Results and Discussion

4.1. The solution Dirac equation for q-deformed trigonometric Scarf and trigonometric tensor potential solution for pseudospin symmetric case

By inserting equations (17) and (18) into equation (12) we obtain

$$\left\{\frac{d^2}{dr^2} - \frac{\kappa(\kappa-1)}{r^2} - \frac{2\kappa}{r} a\left(V_2 \cot_q ar + V_3 \csc_q ar\right) - a^2\left(V_2 \cot_q ar + V_3 \csc_q ar\right)^2 + a^2 V_2 q \csc_q^2 ar\right\} G_{n\kappa}(r)$$
$$+ a^2 V_3 \csc_q ar \cot_q ar G_{n\kappa}(r) + a^2\left(\frac{V_0}{\sin_q^2 ar} - \frac{V_1 \cos_q ar}{\sin_q^2 ar}\right)\left(M - E_{n\kappa} + C_{ps}\right) G_{n\kappa}(r) = \left(M + E_{n\kappa}\right)\left(M - E_{n\kappa} + C_{ps}\right) G_{n\kappa}(r) \tag{41}$$

Equation (38) can not be solved exactly except we use an approximation to the $\frac{1}{r^2}$ term. For small a, $ra \ll 1$ the approximation of $\frac{1}{r^2}$ is given as [18]

$$\frac{1}{r^2} \cong \frac{a^2}{\left(\sin_q^2 ar\right)} \tag{42}$$

Equation (39) is substituted into equation (38) then we get

$$\left\{\frac{d^2}{dr^2}-\frac{a^2\left\{\kappa(\kappa-1)+2\kappa V_3+V_3^2+V_2^2q-V_2q-V_0\gamma_{ps}\right\}}{\sin_q^2 ar}-\frac{a^2(2\kappa V_2+2V_2V_3-V_3+V_1\gamma_{ps})\cos_q ar}{\sin_q^2 ar}\right\}G_{n\kappa}(r) \tag{43}$$
$$=\left\{\left(M+E_{n\kappa}\right)\left(M-E_{n\kappa}+C_{ps}\right)-a^2V_2^2\right\}G_{n\kappa}(r)$$

By setting

$$A_{ps}=\left\{\kappa(\kappa-1)+2\kappa V_3+V_3^2+V_2^2q-V_2q-V_0\gamma_{ps}\right\};\quad \gamma_{ps}=\left(M-E_{n\kappa}+C_{ps}\right) \tag{44}$$

$$B_{ps}=(V_3-2\kappa V_2-2V_2V_3-V_1\gamma_{ps});\quad E'_{ps}=\left\{\frac{\left(M+E_{n\kappa}\right)\left(M-E_{n\kappa}+C_{ps}\right)}{a^2}-V_2^2\right\} \tag{45}$$

in equation (40) then equation (40) reduces to one dimensional Schrodinger-type equation

$$\left\{\frac{d^2}{dr^2}-\frac{a^2A_{ps}}{\sin_q^2 ar}+\frac{a^2B_{ps}\cos_q ar}{\sin_q^2 ar}\right\}G_{n\kappa}(r)=a^2E'_{ps}G_{n\kappa}(r) \tag{46}$$

To simplify the solution of equation (46) we introduce new variable, $\cos_q ar=ix\sqrt{q}$ then equation (46) becomes

$$\left\{\left(1+x^2\right)\frac{\partial^2}{\partial x^2}+x\frac{\partial}{\partial x}+\frac{A_{ps}}{q(1+x^2)}-\frac{B_{ps}i\sqrt{q}x}{q(1+x^2)}+E_{ps}'\right\}G_{n\kappa}(x)=0 \tag{47}$$

According to equation (26) the solution of equation (47) is set to be

$$G_{n\kappa}(x)=g_n(x)=\left(1+x^2\right)^{\frac{\beta}{2}}e^{-\frac{\alpha}{2}\tan^{-1}x}D_n^{(\beta,\alpha)}(x) \tag{48}$$

so from equations (47) and (48) we get

$$\left(1+x^2\right)\frac{\partial^2 D_n^{(\beta,\alpha)}(x)}{\partial x^2}+\left\{x\left(2\beta+1\right)-\alpha\right\}\frac{\partial D_n^{(\beta,\alpha)}(x)}{\partial x}-\left\{\frac{q\beta\alpha x-\frac{q}{2}\alpha x-\frac{q\alpha^2}{4}+q\beta^2-q\beta-A_{ps}+B_{ps}ix\sqrt{q}}{q\left(1+x^2\right)}-E_{ps}'-\beta^2\right\}D_n^{(\beta,\alpha)}(x)=0 \tag{49}$$

By setting

$$q\beta\alpha-\frac{q}{2}\alpha+B_{ps}i\sqrt{q}=0 \quad\text{and}\quad -\frac{q\alpha^2}{4}+q\beta^2-q\beta-A_{ps}=0 \tag{50}$$

in equation (49) we obtain differential equation that satisfies Romanovski polynomials given as

$$\left(1+x^2\right)\frac{\partial^2 D_n^{(\beta,\alpha)}(x)}{\partial x^2}+\left\{x\left(2\beta+1\right)-\alpha\right\}\frac{\partial D_n^{(\beta,\alpha)}(x)}{\partial x}-\left\{-E_{ps}'-\beta^2\right\}D_n^{(\beta,\alpha)}(x)=0 \tag{51}$$

By comparing equations (31) and (51) we get

$$\left(2\beta+1\right)=2(1-p)\rightarrow p=-\beta+1/2;\ \alpha=-q';\ -E_{ps}'-\beta^2=n(n-1)+2n(1-p) \tag{52}$$

By manipulating equations (50) we obtain the values of β and α that have physical meaning are

$$\beta-\frac{1}{2}=\sqrt{\frac{\left(\frac{q}{4}+A_{ps}\right)+B_{ps}\sqrt{q}}{4q}}-\sqrt{\frac{\left(\frac{q}{4}+A_{ps}\right)-B_{ps}\sqrt{q}}{4q}} \tag{53}$$

$$\alpha=-\frac{B_{ps}i}{\sqrt{q}(\beta-1/2)}=-i\omega=-i\left(\sqrt{\frac{\left(\frac{q}{4}+A_{ps}\right)+B_{ps}\sqrt{q}}{q}}+\sqrt{\frac{\left(\frac{q}{4}+A_{ps}\right)-B_{ps}\sqrt{q}}{q}}\right) \tag{54}$$

By manipulating equations (45), (52), and (53) we obtain the relativistic energy equation as

$$\left\{\frac{(M+E_{n\kappa})(E_{n\kappa}-M-C_{ps})}{a^2}+V_2^2\right\}=\left\{\sqrt{\frac{\left(\frac{q}{4}+A_{ps}\right)+B_{ps}\sqrt{q}}{4q}}-\sqrt{\frac{\left(\frac{q}{4}+A_{ps}\right)-B_{ps}\sqrt{q}}{4q}}+n+1/2\right\}^2 \tag{55}$$

The relativistic energy $E_{n\kappa}$ calculated from equation (55) using MatLab 11 programming is presented in Table1. It is shown in Table 1 that there is degeneracy energy for $\left(n,l,j=l+\frac{1}{2};n,l+2,j=l-\frac{1}{2}\right)$ pair, and this degeneracy energy is removed by the presence of the tensor coupling potential. Table 2 shows the relativistic energy as a function of the q-deformed parameter. The system is not deformed for q = 1 and otherwise the system is underlying deformation.

To determine the wave function of the system, firstly we determine the weight function. By using equations (53) and (54) we obtain

$$w^{(p,q')}(x)=(1+x^2)^{\beta-1/2}e^{\frac{B_{ps}i}{\sqrt{q}(\beta-1/2)}\tan^{-1}(x)}=\left(1-\frac{\cos_q ar}{\sqrt{q}}\right)^{\beta-1/2-\omega i/2}\left(1+\frac{\cos_q ar}{\sqrt{q}}\right)^{\beta-1/2+\omega i/2} \tag{56}$$

$$R_n^{(-\beta+1/2,-\alpha)}(x)=D_n^{(\beta,\alpha)}(x)=\frac{1}{(1+x^2)^{\beta-1/2}e^{i\omega\tan^{-1}(x)}}\frac{d^n}{dx^n}\left\{(1+x^2)^{\beta-1/2+n}e^{i\omega\tan^{-1}(x)}\right\} \tag{57}$$

In equation (56) we have applied the trigonometric-hyperbolic relation

$$\tan^{-1}\left(-\frac{i\cos_q ar}{\sqrt{q}}\right)=-i\tanh^{-1}\left(\frac{\cos_q ar}{\sqrt{q}}\right)=-\frac{i}{2}\ln\left(\frac{1+(\cos_q ar)/\sqrt{q}}{1-(\cos_q ar)/\sqrt{q}}\right) \tag{58}$$

The wave function of the system obtained from equations (56) and (57) is given as

$$G_{n\kappa}(x)=g_n(x)=\left(1+x^2\right)^{\frac{\beta}{2}}e^{\frac{i\omega}{2}\tan^{-1}x}D_n^{(\beta,\alpha)}(x)=\left(1+x^2\right)^{\frac{\beta}{2}}e^{\frac{\omega i}{2}\tan^{-1}x}R_n^{(-\beta+1/2,-\alpha)}(x) \tag{59}$$

The ground state wave function of lower component of Dirac spinor is obtained from equations (58) and (59), when for $n = 0$ the lowest degree of Romanovski polynomials, $R_0^{(-\beta+1/2,-\alpha)}(x)=1$, so

$$G_{0\kappa}(x)=G_{0\kappa}(r)=C\left(1-\frac{\cos_q ar}{\sqrt{q}}\right)^{\beta-\omega i/2}\left(1+\frac{\cos_q ar}{\sqrt{q}}\right)^{\beta+\omega i/2} \tag{60}$$

Table 1. The relativistic energy $E_{n\kappa}$ with different values of l and n for V_0=4 fm^{-1}; V_1=3 fm^{-1}; a = 0.05 fm^{-1}; M =3 fm^{-1}; q = 1; and C_{ps} = -5 fm^{-1}

l	n	κ	$n,l,$ $j=l+\frac{1}{2}$	$E_{n\kappa}$, V_2, V_3 = 0	$E_{n\kappa}$ V_2=0.6 V_3 =0.8	n	κ	$n,l+2,$ $j=l-\frac{1}{2}$	$E_{n\kappa}$ V_2, V_3 = 0	$E_{n\kappa}$ V_2=0.6 V_3 =0.8
0	0	-1	$0s_{1/2}$	-1.989967362	-1.996787442	0	2	$0d_{3/2}$	-1.989967362	-1.980582275
1	0	-2	$0p_{3/2}$	-1.977729417	-1.988336363	0	3	$0f_{5/2}$	-1.977729417	-1.964869912
2	0	-3	$0d_{5/2}$	-1.961071492	-1.975177161	0	4	$0g_{7/2}$	-1.961071492	-1.944995326
3	0	-4	$0f_{7/2}$	-1.940392072	-1.957672974	0	5	$0h_{9/2}$	-1.940392072	-1.921420969
0	1	-1	$1s_{1/2}$	-1.977559362	-1.987684954	1	2	$1d_{3/2}$	-1.977559362	-1.964271652
1	1	-2	$1p_{3/2}$	-1.960906399	-1.974854327	1	3	$1f_{5/2}$	-1.960906399	-1.944567394
2	1	-3	$1d_{5/2}$	-1.940222876	-1.957406381	1	4	$1g_{7/2}$	-1.940222876	-1.921067297
3	1	-4	$1f_{7/2}$	-1.915947799	-1.935988592	1	5	$1h_{9/2}$	-1.915947799	-1.894278609

The upper component of Dirac spinor for pseudospin symmetry obtained using equations (8) and (59) is

$$F_{n\kappa}(r)=\frac{\left(\frac{d}{dr}-\frac{\kappa}{r}+U(r)\right)}{\left(M-E_{n\kappa}+C_{ps}\right)}G_{n\kappa}(r)$$

For exact pseudospin symmetric case when $C_{ps}=0$, the upper spinor exist if $M\neq E_{n\kappa}$ which means that the relativistic energies for pseudospin symmetric case are always negative.

4.2. The solution Dirac equation for q-deformed trigonometric Scarf and trigonometric tensor potential solution for spin symmetric case

By inserting equations (17) and (18) into equation (12) we obtain

$$\left\{\frac{d^2}{dr^2}-\frac{\kappa(\kappa+1)}{r^2}-\frac{2\kappa}{r}a\left(V_2\cot_q ar+V_3\csc_q ar\right)-a^2\left(V_2\cot_q ar+V_3\csc_q ar\right)^2-a^2V_2q\csc_q^2 ar\right\}F_{n\kappa}(r)$$
$$-a^2V_3\csc_q ar\cot_q arF_{n\kappa}(r)-a^2\left(\frac{V_0}{\sin_q^2 ar}-\frac{V_1\cos_q ar}{\sin_q^2 ar}\right)(M+E_{n\kappa}-C_s)F_{n\kappa}(r)=(M-E_{n\kappa})(M+E_{n\kappa}-C_s)F_{n\kappa}(r) \quad (62)$$

The solution steps of Dirac equation for spin symmetric case is similar with the solution steps for pseudospin symmetric case, therefore by using the approximation scheme expressed in equation (39) into equation (61) we get

$$\left\{\frac{d^2}{dr^2}-\frac{a^2\left\{\kappa(\kappa+1)+2\kappa V_3+V_3^2+V_2^2q-V_2q+V_0\gamma_s\right\}}{\sin_q^2 ar}-\frac{a^2(2\kappa V_2+2V_2V_3-V_3-V_1\gamma_{ps})\cos_q ar}{\sin_q^2 ar}\right\}F_{n\kappa}(r) \quad (63)$$
$$=\left\{(M-E_{n\kappa})(M+E_{n\kappa}-C_s)-a^2V_2^2\right\}F_{n\kappa}(r)$$

By setting

$$A_s=\left\{\kappa(\kappa+1)+2\kappa V_3+V_3^2+V_2^2q-V_2q+V_0\gamma_s\right\};\ \gamma_s=\left(M+E_{n\kappa}-C_s\right) \quad (64)$$

$$B_s=(V_3-2\kappa V_2-2V_2V_3+V_1\gamma_s);\ E'_s=\left\{\frac{(M-E_{n\kappa})(M+E_{n\kappa}-C_s)}{a^2}-V_2^2\right\} \quad (65)$$

in equation (63) then equation (63) reduces to one dimensional Schrodinger-type equation

$$\left\{\frac{d^2}{dr^2}-\frac{a^2A_s}{\sin_q^2 ar}+\frac{a^2B_s\cos_q ar}{\sin_q^2 ar}\right\}F_{n\kappa}(r)=a^2E_s'F_{n\kappa}(r) \tag{66}$$

By variable substitution given as $\cos_q ar = ix\sqrt{q}$ in equation (66) we obtain

$$\left\{\left(1+x^2\right)\frac{\partial^2}{\partial x^2}+x\frac{\partial}{\partial x}+\frac{A_s}{q(1+x^2)}-\frac{B_s i\sqrt{q}x}{q(1+x^2)}+E_s'\right\}F_{n\kappa}(x)=0 \tag{67}$$

The solution of equation (67) is obtained by setting

$$F_{n\kappa}(x)=g_n(x)=\left(1+x^2\right)^{\frac{\beta}{2}}e^{-\frac{\alpha}{2}\tan^{-1}x}D_n^{(\beta,\alpha)}(x) \tag{68}$$

and equation (67) becomes

$$\left(1+x^2\right)\frac{\partial^2 D_n^{(\beta,\alpha)}(x)}{\partial x^2}+\left\{x(2\beta+1)-\alpha\right\}\frac{\partial D_n^{(\beta,\alpha)}(x)}{\partial x}-\left\{\frac{q\beta\alpha x-\frac{q}{2}\alpha x-\frac{q\alpha^2}{4}+q\beta^2-q\beta-A_s+B_s ix\sqrt{q}}{q\left(1+x^2\right)}-E_s'-\beta^2\right\}D_n^{(\beta,\alpha)}(x)=0 \tag{69}$$

By setting

$$q\beta\alpha-\frac{q}{2}\alpha+B_s i\sqrt{q}=0 \qquad -\frac{q\alpha^2}{4}+q\beta^2-q\beta-A_s=0 \tag{70}$$

in equation (69) we get

$$\left(1+x^2\right)\frac{\partial^2 D_n^{(\beta,\alpha)}(x)}{\partial x^2}+\left\{x(2\beta+1)-\alpha\right\}\frac{\partial D_n^{(\beta,\alpha)}(x)}{\partial x}-\left\{-E_s'-\beta^2\right\}D_n^{(\beta,\alpha)}(x)=0 \tag{71}$$

By comparing equations (26) and (71) we get

$$(2\beta+1)=2(1-p)\rightarrow p=-\beta+1/2;\ \alpha=-q';\ \text{and}\ -E_s'-\beta^2=n(n-1)+2n(1-p) \tag{72}$$

The values of β and α that have physical meaning are

$$(\beta_s-1/2)=\sqrt{\frac{\left(\frac{q}{4}+A_s\right)+B_s\sqrt{q}}{4q}}-\sqrt{\frac{\left(\frac{q}{4}+A_s\right)-B_s\sqrt{q}}{4q}}$$

$$\text{and } \alpha_s=-i\omega_s=-i\sqrt{\frac{\left(\frac{q}{4}+A_s\right)+B_s\sqrt{q}}{q}}-\sqrt{\frac{\left(\frac{q}{4}+A_s\right)-B_s\sqrt{q}}{q}} \tag{73}$$

By using equations (71) and (73) we have

$$\left\{\frac{\left(E_{n\kappa}-M\right)\left(E_{n\kappa}+M-C_s\right)}{a^2}+V_2^2\right\}=\left\{\sqrt{\frac{\left(\frac{q}{4}+A_s\right)+B_s\sqrt{q}}{4q}}-\sqrt{\frac{\left(\frac{q}{4}+A_s\right)-B_s\sqrt{q}}{4q}}+n+1/2\right\}^2 \tag{74}$$

By using equations (33), (58) and (73) we obtain the weight function given as

$$w^{(p,q')}(x)=(1+x^2)^{-p}e^{q'\tan^{-1}(x)}=\left(1-\frac{\cos_q ar}{\sqrt{q}}\right)^{\beta_s-1/2-\omega_s/2}\left(1+\frac{\cos_q ar}{\sqrt{q}}\right)^{\beta_s-1/2+\omega_s/2} \tag{75}$$

and the corresponding Romanovski polynomials is

$$R_n^{(\beta_s-1/2,-\alpha_s)}(x) = D_n^{(\beta,\alpha)}(x) = \frac{1}{(1+x^2)^{\beta_s-1/2}e^{i\omega_s\tan^{-1}(x)}}\frac{d^n}{dx^n}\left\{(1+x^2)^{\beta_s-1/2+n}e^{i\omega_s\tan^{-1}(x)}\right\} \quad (76)$$

The wave function of the system obtained from equations () and () is given as

$$F_{n\kappa}(x) = g_n(x) = \left(1+x^2\right)^{\frac{\beta_s}{2}} e^{-\frac{\alpha_s}{2}\tan^{-1}x} D_n^{(\beta,\alpha)}(x) = \left(1+x^2\right)^{\frac{\beta_s}{2}} e^{-i\frac{\omega_s}{2}\tan^{-1}x} R_n^{(\beta_s-1/2,-\alpha_s)}(x) \quad (77)$$

The lower component of Dirac spinor for symmetric case is obtained using equations (7) and (77) as

$$G_{n\kappa}(r) = \frac{\left(\frac{d}{dr}+\frac{\kappa}{r}-U(r)\right)}{\left(M+E_{n\kappa}-C_{ps}\right)}F_{n\kappa}(r) \quad (78)$$

Exact spin symmetry is special case for spin symmetry when $C_{ps}=0$, therefore the lower spinor component exist only if $M \neq -E_{n\kappa}$ which means that the relativistic energy, $E_{n\kappa}$, is always positive.

5. Conclusion

The Dirac equation for q-deformed trigonometric Scarf potential with q-deformed trigonometric cotangent and cosecant tensor potential in the scheme of centrifugal term approximation is exactly solved using Romanovski polynomials both for pseudospin and spin symmetric cases. The relativistic energy spectra calculated using Matlab are always positive for spin symmetric case and negative for pseudospin symmetric case. The lower and upper component of Dirac spinors are obtained in the form of Romanovski polynomial. The tensor potential causes the degeneracy energy is removed both for pseudospin and spin symmetries.

Acknowledgement
This work is partly funded by Hibah Pascasarjana DIKTI, Sebelas Maret University (UNS), No.165/UN27.11/PN2013.

Reference

[1] Hamzavi M and Rajabi AA 2013 *Adv. High. En. Phys.* **2013** 196986.
[2] Ikhdair SM and Sever R 2010 *arXiv*: 1001.4327v2.
[3] Eshghi M 2011 *J. Sci. Res.* **3**/3 493-500.
[4] Ikot AN 2013 *Commun. Theor Phys.* **59** 268-272.
[5] Aydogdu O, Maghsoodhi E, and Hassanabadi H 2013 *Chin Phys. B* **22** 010302-1-9.
[6] Ai-Ping Z and Wen-Chao Q 2007 *High Energy Phys. and Nucl. Phys.* **31** 1027-1031.
[7] Chen CH 2005 *Phys. Lett. A* **339** 283-287.
[8] Aydogdu O and Sever R 2011 *Phys. Lett. B.* **703** 379-385.
[9] Soylu A, Bayrak O, and Boztosun I 2007 *J. Math. Phys.* **48** 082302.
[10] Zhang LH, Li XP, and Jia CS 2008 *Phys. Lett. A* **372** 2201-2207.
[11] Wei GF and Dong SH 2008 *Phys. Lett. A* **373** 49-53.
[12] Ikhdair SM and Sever R 2010 *Appl. Math. Comput.* **216** 911-923.
[13] Hamzavi M, Rajabi AA, and Hassanabadi H 2010 *Phys. Lett.* A **374** 4303 - 4307.
[14] Wei GF and Dong SH 2008 *Phys. Lett. A* **373** 49 - 53.
[15] Xu Y, He S, and Jia CS 2008 *J. Phys. A: Math. and Theor.* **41** 0255302-1 - 0255302-8.
[16] Greene RL and Aldrich C 1976 *Physical Review A* **14**(6) 2363–2366.
[17] Hecht KT and Adler A 1969 *Nucl. Phys. A* **137** 129.
[18] Arima A, Harvey M, Shimizu K 1969 *Phys. Lett. B* **30**(8) 517-522.
[19] Ginocchio JN 2004 *Phys. Rep. C* **69** 034313.
[20] Cari C, Suparmi A, Deta UA, SW Intan 2013 *Makara J. Sci.* **17** 3 in press.
[21] Meng J, Sugaware-Tanabe K, Yamaji S, Arima A 1999 *Phys. Rev. C* **59** 154.

[22] Suparmi A and Cari C 2013 *At. Indonesia J.* **39** 3 in press.
[23] Troltenier D, Bahri C, and Draayer JP 1995 *Nucl. Phys. A* **586**(1) 53-72.
[24] Page PR, Goldman T, and Ginocchio JN 2001 *Phys. Rev. Lett.* **86** (2-8) 204-207.
[25] Ikhdair SM, Falaye BJ, and Adepoju AG 2013 *arXiv*: 1308.0155v [quant-ph].
[26] Ginocchio JN 2005 *Phys. Rep.* **414** 165-262.
[27] Ginocchio JN and Madland DG 1998 *Phys. Rev. C* **57**(3) 1167 - 1173.
[28] Alberto P, Fiolhais M, Malheiro M, Delfino A, and Chiapparini M 2002 *Phys. Rev. C* **6** 034307-1 - 034307-9.
[29] Troltenier D, Nazarewicz W, Szymanski Z, and Draayer JP 1994 *Nucl. Phys. A.* **67** 591-610.
[30] Stuchbery AE 2002 *Nucl. Phys. A* **700** 83-116.
[31] Aydogdu O and Sever R 2010 *Few-Body system* **47** 193 – 200.
[32] Akcay H 2009 *Phys. Lett. A* **373** 616 – 620.
[33] Hamzavi M, Rajabi AA, and Hassanabadi H 2010 *Few-Body Sys.* **48** 171-183.
[34] Koepf W and Masjed-Jamei M 2006 *Integral Transform and Special Functions* **17** 559.
[35] Routh EJ 1884 *Proc. London Math. Soc.* **16** 245.
[36] Romanovski VI 1929 *Compt. Rend. Ac. Sci. Paris* **188** 1023.
[37] Alvarez-Castillo DE and Kirbach M 2007 *Rev. Mex. Fis* **E 53**(2) 143-154.
[38] Raposo A, Weber HJ, Alvarez-Castillo DE, and Kirchbach M 2007 *C. Eur. Phys. J.* **5** 253.
[39] Suparmi A, Cari C, Handhika J, Yanuarief C, and Marini H 2012 *IOSR, J. App. Phys.* **2** (2) 43-51.
[40] Alvarez-Castillo DE, Compean CB, and Kirbach M 2011 *arXiv* 1105. 1354v1 [quant-ph].
[41] Suparmi A, Cari C, Handhika J 2012 *J. Phys. Conf. Ser.* **423** 012039.
[42] Suparmi A and Cari C 2013 *The 4th ICANCE 2013 September16-19*, Bali, Indonesia.
[43] [44] Arai A 1991 *J. Math. Anal. Apply.* **158** (63) 63-79.
[44] [45] Dutra AS 2005 *arXiv*:quant-ph/0501094v1.
[45] [46] Suparmi A, Cari C, Yuliani H 2013 *Ad. Phys. Theor. Appl., IISTE* **16** 64-74.
[46] [47] Nikivorof AF and Uvarov VB 1988 *Special Functions of Mathematical. Physics – A Unified Introduction with Applications* (Germany: Birkhauser Verlag Basel) pp. 317-318.

Theoretical study of diffusion processes around a non-rotating neutron star

D. Andra[1,2], **M.F. Rosyid**[1]

[1]Jurusan Fisika FMIPA Universitas Gadjah Mada, Yogyakarta, Indonesia
[2]Jurusan Pendidikan MIPA FKIP Universitas Lampung, Bandar Lampung, Indonesia

E-mail: doniandra.fisika10@gmail.com, farchani@ugm.ac.id

Abstract. The general relativistic diffusion process on curved space-time manifold around a non-rotating neutron star has been analyzed. The general relativistic diffusion equation of diffusive particles around non-rotating neutron star is derived by constructing phase space in the parametrization of observer time in the hyperbolic coordinate system. This diffusion equation describes the stochastic dynamic of particles around non-rotating neutron stars. In this work we also have studied the diffusion processes around a non-rotating neutron star for asymptotic case.

1. Introduction

Diffusion processes constitute a class of stochastic processes which are characterized by two properties: the Markovian property and the continuity of trajectories. Diffusion process is a solution to a stochastic differential equation. It is a continuous-time Markov process with almost surely continuous sample paths. Some examples of diffusion processes are Brownian motion, reflected Brownian motion and Ornstein-Uhlenbeck processes. A sample path of a diffusion process models the trajectory of a particle embedded in a flowing fluid and subjected to random displacements due to collisions with molecules, which is called Brownian motion. The position of the particle is then random; its probability density function as a function of space and time is governed by an advection-diffusion equation.

The most fundamental diffusion process is Brownian motion or Wiener process. The Brownian motion process was studied originally by English botanist Robert Brown for modeling the motion of a small particle immersed in a liquid subject to molecular collisions. A mathematical model of the process was first derived by Einstein and the underlying theory was subsequently perfected by Fokker, Planck, and Wiener, among others. Furthermore, the Brownian motion as a diffusion process has developed rapidly both physically and mathematically. Research on Brownian motion is also experiencing growth both relativistic and nonrelativistic.

One study of relativistic diffusion process has been performed by Dunkel and Hanggi. In 2005 Dunkel and Hanggi did research on relativistic Brownian motion for 1+1-dimension case [4]. Then, on the same year they did research on relativistic Brownian motion for 3+1-dimension case [5]. Another researcher who studied about relativistic diffusion process is Joachim Herrmann. In 2009 Herrmann studied about diffusion in the special theory of relativity [7]. One year later, he did research about diffusion in the general theory of relativity [8].

Based on the observations made by astronomers, it has been found that the gas particles which moving toward the compact stars, called as accretion disc. The movement of the particles is determined by the curvature of space-time [2]. Movement of gas particles can be viewed as a process of diffusion. To understand the movement of particles which experiencing diffusion around the compact stars or massive objects, it takes a special review of diffusion processes that meet the rules of Einstein's theory of relativity. That theory give an explanation of the influence of the mass and energy of the curvature of space-time. Higher energy density will cause greater curvature of space-time and vice versa. In other words, the curvature of space-time around objects that have a very high density, for example in the compact stars like white dwarfs, neutron stars, and black holes; is also become very large [1].

In this research we will discuss about diffusion process around non-rotating neutron star. Those study refers to research done by Herrmann [8]. Herrmann has constructed diffusion equation on the general relativity framework. Then, he applied those equation to describe diffusion process in expanding universe. Whereas, in this research we use diffusion equation from Herrmann's work to describe diffusion process around a non-rotating neutron star.

2. Diffusion process

Diffusion processes constitute a class of stochastic processes which are characterized by two properties: the Markovian property and the continuity of trajectories. Now, we give a formal definition of diffusion processes. Let S be a topological space. We sometimes find it convenient to attach an extra point Δ to S, either as an isolated point or as point at infinity if S is locally compact. Thus we set $S' = S \cup \Delta$. The parameter Δ and S or S' is called the terminal point and the state space, respectively. Let $\bar{W}(S)$ be the set of all functions $w : [0, \infty) \ni t \mapsto w(t) \in S'$ such that there exist $0 \leq \zeta(w) \leq \infty$ with the following properties [9]:

- $w(t) \in S$ for all $t \in [0, \zeta(w))$ and the mapping $t \in [0, \zeta(w) \mapsto w(t)$ is continuous,
- $w(t) = \Delta$ for all $t \geq \zeta(w)$.

The parameter $\zeta(w)$ is called the lifetime of the trajectory w. For convenience, we set $w(\infty) = \Delta$ for every $w \in \bar{W}(S)$. A Borel cylinder set in $\bar{W}(S)$ is defined for some integer n, a sequence $0 \leq t_1 < t_2 < ... < t_n$ and a Borel subset A in $S'^n = S' \times S' \times ... \times S'$ as $\pi^{-1}_{t_1,t_2,...,t_n}(A)$. Whereas $\pi_{t_1,t_2,...,t_n} : \bar{W}(S) \mapsto S'^n$ is given by $\pi_{t_1,t_2,...,t_n}(w) = (w(t_1), w(t_2), ..., w(t_n))$.

A Borel subset in a topological space is any set in the smallest σ-field containing all open sets. Let $\mathbb{B}(\bar{W}(S))$ be the σ-field in $\bar{W}(S)$ generated by all Borel cylinder sets and let $\mathbb{B}_t(\bar{W}(S))$ be that generated by all cylinder sets up to time t, i.e, sets expressed in the form $\pi^{-1}_{t_1,t_2,...,t_n}(A)$ where $t_n \leq t$. A family of probabilities $P_x, x \in S'$ on $(\bar{W}(S), \mathbb{B}_t(\bar{W}(S)))$ is called Markovian if

- $P_x w; w(0) = x = 1$ for every $x \in S'$,
- $x \in S \mapsto P_x(A)$ is Borel measurable for each $A \in \mathbb{B}(\bar{W}(S))$, and.
- for every $t > s \geq 0, A \in \mathbb{B}_s(\bar{W}(S))$ and a Borel subset Γ in S',

$$P_x(A \cap w; w(t) \in \gamma) = \int_A P_{w^t(s)}[w; w(t-s) \in \Gamma] P_x(dw')$$

for every $x \in S'$.

Definition 1 *[9] A family of probabilities $P_{x x \in S'}$ on $(\bar{W}(S), \mathbb{B}_t(\bar{W}(S)))$ is called a system of diffusion measures, or simply a diffusion if it is a strongly Markovian system.*

Definition 2 *[9] A stochastic process $X = X(t)$ on S' defined on a propability space $(\Omega, \mathbb{F}, P)$ is called a diffusion process on S if there exists a system of diffusion measures $P_{x x \in S'}$, such that, for almost all ω, $[t \mapsto X(t)] \in \bar{W}(S)$ and the probability law on $\bar{W}(S)$ of $[t \mapsto X(t)]$ coincides with $P_\mu(\ldots\ldots\ldots) = \int_{S'} P_x(\ldots\ldots\ldots)\mu(dx)$ where μ is the Borel measure on S' defined by $\mu(dx) = P\omega; X(0, \omega) \in dx$ and is called the initial distribution of X.*

Journal of Physics: Conference Series **539** (2014) 012005 doi:10.1088/1742-6596/539/1/012005

If $X = (X(t))$ is a diffusion process and if we set $\zeta(\omega) = inft; X(t) = \Delta$, then it is clear that, with probability one, $(0, \zeta] \ni t \mapsto X(t) \in S$ is continous and $X(t) = \Delta$ for all $t \leq \zeta$. The parameter ζ is called the life time of the diffusion process X. A stochastic process X is called conservative if $\zeta(\omega) = \infty$.

2.1. Brownian motion

Brownian motion is the most fundamental of diffusion process. Brownian motion is a Markov process with a continuous state space and a continuous index set. Brownian motion is closely linked to the normal distribution. Recall that a random variable X is normally distributed with mean μ and variance σ^2 if [11]

$$\mathbb{P}\{X > x\} = \frac{1}{\sqrt{2\pi\sigma^2}} \int_x^{\infty} e^{-\frac{(u-\mu)^2}{2\sigma^2}} du, \quad \text{untuk} \quad x \in \mathbb{R}$$

Definition 3 *A real-valued stochastic process $\{B(t) : t \geq 0\}$ is called a (linear) Brownian motion with start in $x \in \mathbb{R}$ if the following holds [11]:*

- $B(0) = x$,
- *the process has independent increments, i.e. for all times $0 \leq t_1 \leq t_2 \leq ... \leq t_n$ the increments $B(t_n) - B(t_{n-1}), B(t_{n-1}) - B(t_{n-2}), ..., B(t_2) - B(t_1)$ are independent random variables,*
- *for all $t \geq 0$ and $h > 0$, the increments $B(t + h) - B(t)$ are normally distributed with expectation zero and variance h,*
- *almost surely, the function $t \longmapsto B(t)$ is continuous.*

2.2. Langevin equation

Langevin equation is a differential stochastic equation that describes time evolution an observable. Consider a Brownian particle with rest mass m, proper time τ, and 4-velocity $u^\beta(\tau)$. The 4-momentum of the particle is given by $p^\alpha = mu^\alpha$, where $u_\alpha u^\alpha \equiv -c^2$. Assume that the particle is surrounded by an isotropic, homogeneous heat bath with constant 4-velocity U^β and, additionally, subject to an external 4-force $K^\alpha(x^\nu, p^\mu)$ such as, e.g., the Lorentz force. Then, the relativistic Langevin equations of motion read [5]

$$dx^\alpha(\tau) = \frac{p^\alpha}{m} d\tau \tag{1}$$

$$dp^\alpha(\tau) = K^\alpha - \kappa^\alpha_\beta [p^\beta - mU^\beta] d\tau + w^\alpha(\tau) \tag{2}$$

For an isotropic homogeneous heat bath, the friction tensor κ^α_β is given by [5]

$$\kappa^\alpha_\beta = \kappa \left(\eta^\alpha_\beta + \frac{u^\alpha u_\beta}{c^2} \right) \tag{3}$$

with κ denoting the scalar viscous friction coefficient measured in the rest frame of the particle. Furthermore, the relativistic Wiener increments $w^\alpha(\tau) \equiv dW^\alpha(\tau)$ are distributed according to the probability density [5]

$$\mathcal{P}^{1+3}[w^\alpha(\tau)] = \frac{c}{(4\pi D d\tau)^{3/2}} exp\left[-\frac{w_\alpha(\tau) w^\alpha(\tau)}{4Dd\tau} \right] \text{x} \delta [u_\alpha w^\alpha(\tau)] \tag{4}$$

where D is the scalar noise amplitude parameter measured in the rest frame of the particle.

2.3. Fokker-Planck equation
The Fokker-Planck equation is a special type of master equation, which is often used as an approximation to the actual equation or as a model for more general Markov processes. The Fokker-Planck equation is a master equation in which $\mathbb{W}$ is a differential operator of second order [12]

$$\frac{\partial P(y,t)}{\partial t} = -\frac{\partial}{\partial y}A(y)P + \frac{1}{2}\frac{\partial^2}{\partial y^2}B(y)P. \tag{5}$$

Those equation is also called Smoluchowski equation, generalized diffusion equation or second Kolmogorov equation. The first term on the right-hand side is called transport term, convection term, or drift term; the second one diffusion term or fluctuation term. The range of y is necessarily continuous, and for the present time is supposed to be $(-\infty, +\infty)$. The coefficients $A(y)$ and $B(y)$ may be any real differentiate functions with the sole restriction $B(y) > 0$. The equation can be broken up into a continuity equation for the probability density

$$\frac{\partial P(y,t)}{\partial t} = -\frac{\partial J(y,t)}{\partial y}, \tag{6}$$

where $J(y,t)$ is the probability flux, and a constitutive equation

$$J(y,t) = A(y)P - \frac{1}{2}\frac{\partial}{\partial y}B(y)P. \tag{7}$$

Let $P(y,t|y_1,t_1)$ for $t \geq t_1$ be that solution of equation (5) that at t_1 reduces to $\delta(y - y_1)$. Then, we can construct a Markov process whose transition probability is $P(y_2,t_2|y_1,t_1)$ and whose one-time distribution $P_1(y_1,t_1)$ may still be chosen arbitrarily at one initial time t_0. If we choose for P_1 the stationary solution of (5),

$$P^s(y) = \frac{const.}{B(y)}exp\left[2\int_0^y \frac{A(y')}{B(y')}dy'\right], \tag{8}$$

the resulting Markov process is stationary. This is only possible, however, when this P^s is integrable so that it can be normalized to represent a probability distribution.

3. Neutron star

Neutron stars play a unique role in physics and astrophysics. On the one hand, they contain matter under extreme physical conditions, and their theories are based on risky and far extrapolations of what we consider reliable physical theories of the structure of matter tested in laboratory. On the other hand, their observations offer the unique opportunity to test these theories. Moreover, neutron stars are important dramatic personae on the stage of modern astrophysics; they participate in many astronomical phenomena. Neutron stars contain the matter of density ranging from a few $\mathtt{gcm}^{-3}$ at their surface, where the pressure is small, to more than $10^{15}\mathtt{gcm}^{-3}$ at the center, where the pressure exceeds $10^{36}\mathtt{dyn\ cm}^2$. To calculate neutron star structure, one needs the dependence of the pressure on density, the so called equation of state (EOS), in this huge density range, taking due account of temperature, more than 10^9 K in young neutron stars, and magnetic fields, sometimes above 10^{15} G [6].

Neutron stars are compact stars which contain matter of supranuclear density in their interiors. They have typical masses $M \sim 1,4M\odot$,where $M\odot$ is the Solar mass, and radii $R \sim 10$ km. Thus, their masses are close to the solar mass $M\odot = 1.989 \times 10^{33}$g, but their radii are 10^5 times smaller than the solar radius $R\odot = 6.96 \times 10^5$km. Because of its small size and high density, a neutron star possesses a surface gravitational field about 2×10^{11}

times that of Earth. Accordingly, neutron stars possess an enormous gravitational energy $E_{grav} \sim GM^2/R \sim 5\times10^{53}\mathtt{erg} \sim 0.2Mc^2$ and surface gravity $g \sim GM/R^2 \sim 2\times10^{14}\mathtt{cm\ s}^{-2}$, where G is the gravitational constant and c is the speed of light. Clearly, neutron stars are very dense. Their mean mass density is $\bar{\rho} \simeq 3M/4\pi R^3 \simeq 7\times10^{14} gcm^{-3} \sim (2-3)\rho_0$, where $\rho_0 = 2.8\times10^{14}\mathtt{gcm}^{-3}$ is is the so called normal nuclear density, the mass density of nucleon matter in heavy atomic nuclei. The central density of neutron stars is even larger, reaching $(1020)\rho_0$. By all means, neutron stars are the most compact stars known in the Universe [6].

A neutron star can be subdivided into the atmosphere and four main internal regions: the outer crust, the inner crust, the outer core, and the inner core. The atmosphere is a thin plasma layer, where the spectrum of thermal electromagnetic neutron star radiation is formed. The outer crust (the outer envelope) extends from the atmosphere bottom to the layer of the density $\rho \approx 4\times10^{11}\mathtt{gcm}^{-3}$. Its thickness is some hundred meters. Its matter consists of ions Z and electrons e. The inner crust (the inner envelope) may be about one kilometer thick. The density ρ in the inner crust varies from ρ_{ND} at the upper boundary to $0{,}5\rho_0$ at the base. The matter of the inner crust consists of electrons, free neutrons, and neutron-rich atomic nuclei. The outer core occupies the density range $0,5\rho_0 \le \rho \le 2\rho_0$ and is several kilometers thick. Its matter consists of neutrons with several per cent admixture of protons, electrons, and possibly muons. The inner core, where $\ge 2\rho_0$, occupies the central regions of massive neutron stars. Its radius can reach several kilometers, and its central density can be as high as (10-50)ρ_0. Its composition and the EOS are very model dependent. Several hypotheses have been put forward, predicting the appearance of new fermions and/or boson condensates [6].

4. Diffusion process around a neutron star

In this section we will discuss about relativistic diffusion process around a non-rotating neutron star. We will derive sthochastic differential equation which describes the diffusive particles dynamic around a non-rotating neutron star. The derivation of those equation refers to Herrmann's work [8]. Herrmann has generalized the Markovian diffusion theory in the phase space within the general theory of relativity framework. He has derived the general relativistic Kramers equation in the phase space both in the parametrization of phase space proper time and the coordinate time. The general relativistic Kramers equation in the parametrization of observer time x_0 derived by Herrmann is given by

$$N^{-1}v^0\frac{\partial}{\partial x^0}\Phi \;=\; -v^M div_x(e_M\Phi) - div_v(F\Phi) + \frac{D}{2}\Delta_v\Phi, \qquad (9)$$

where $N = (g_{00})^{-1/2}$, $div_x(e_M(\mathbf{x})\Phi)$ is the divergence operator in the position space, $div_v(\mathbf{F}\Phi)$ is the divergence operator in the position space, and Δ_v is the Laplace-Beltrami operator. The divergence operator in the position space, the divergence operator in the position space, and the Laplace-Beltrami operator are expressed in the hyperbolic coordinate system. Then, this equation is used to construct the diffusion equation to describe the diffusive particles dynamic around a non-rotating neutron star.

The metric of nonrotating neutron star is given by

$$ds^2 = c^2e^{2\nu}dt^2 - e^{2\lambda}dr^2 - r^2(d\theta^2 + \sin^2\theta d\phi^2), \qquad (10)$$

where ν and λ are a function with parameter r. The metric coefficients are $g_{00} = e^{2\nu}, g_{11} = -e^{2\lambda}, g_{22} = -r^2, g_{33} = -r^2\sin^2\theta$, and the others are zero. Then, the Christoffel symbols for this metric are given by

$$\begin{aligned}
\Gamma^1_{00} &= \frac{\partial \nu}{\partial r} e^{2\nu - 2\lambda} = \nu' e^{2\nu-2\lambda} \\
\Gamma^1_{11} &= \frac{\partial \lambda}{\partial r} = \lambda' \\
\Gamma^1_{22} &= -re^{-2\lambda} \\
\Gamma^1_{33} &= -r\sin^2\theta e^{-2\lambda} \\
\Gamma^2_{33} &= -\sin\theta\cos\theta \\
\Gamma^0_{01} &= \Gamma^0_{10} = \frac{\partial \nu}{\partial r} = \nu' \\
\Gamma^2_{21} &= \Gamma^2_{12} = \Gamma^3_{31} = \Gamma^3_{13} = \frac{1}{r} \\
\Gamma^3_{32} &= \Gamma^3_{23} = \cot\theta,
\end{aligned} \tag{11}$$

and the others are 0.

Basis of orthonormal frame vectors are determined by

$$\eta^{MN} e^{\mu}_{M}(x) e^{\nu}_{N}(x) = g^{\mu\nu}, \tag{12}$$

Based on the equation of (12) covariant basis of orthonormal frame vectors are given by

$$\begin{aligned}
\epsilon^0_0 &= \sqrt{g^{00}} = \sqrt{e^{2\nu}} = e^{\nu} \\
\epsilon^1_1 &= \sqrt{g^{11}} = \sqrt{e^{2\lambda}} = e^{\lambda} \\
\epsilon^2_2 &= \sqrt{g^{22}} = \sqrt{r^2} = r \\
\epsilon^3_3 &= \sqrt{g^{33}} = \sqrt{r^2\sin^2\theta} = r\sin\theta,
\end{aligned} \tag{13}$$

whereas the contravariant basis of orthonormal frame vectors are given by

$$\begin{aligned}
e^0_0 &= e^{-\nu}; \qquad & e^1_1 = e^{-\lambda} \\
e^2_2 &= \frac{1}{r}; \qquad & e^3_3 = \frac{1}{r\sin\theta}.
\end{aligned} \tag{14}$$

Spin connection coefficients are determined by the equation

$$\Omega^M_{\mu N}(x) = \epsilon^M_{\nu}(x)\Gamma^{\nu}_{\mu\rho} e^{\rho}_{N}(x) + \epsilon^M_{\nu}(x)\partial_{\mu} e^{\nu}_{N}(x). \tag{15}$$

By substituting Christoffel symbols and basis to the equation (15) then we get the spin connection coefficients below

$$\begin{aligned}
\Omega^1_{00} &= \nu' e^{\nu-\lambda}; \qquad & \Omega^0_{01} = \nu' e^{\nu-\lambda} \\
\Omega^2_{21} &= e^{-\lambda}; \qquad & \Omega^1_{22} = -e^{-\lambda} \\
\Omega^3_{31} &= e^{-\lambda}\sin\theta; \qquad & \Omega^3_{32} = \cos\theta \\
\Omega^1_{33} &= -e^{-\lambda}\sin\theta; \qquad & \Omega^2_{33} = -\cos\theta,
\end{aligned} \tag{16}$$

and the others are zero.

The forces of curved space-time are calculated by equation

$$F_g^a = -\Omega^a_{\mu B}(x) e^{\mu}_C(x) u^B u^C, \tag{17}$$

where $\Omega^a_{\mu B}$ is the spin connection coefficients ; e^{μ}_C is basis of orthonormal frame vectors; u^B and u^C are 4-velocity in the hyperbolic coordinate system. The 4-velocity in the hyperbolic coordinate system are given by

$$\begin{aligned} u^0 &= \cosh\alpha \\ u^1 &= \sinh\alpha \sin\beta \cos\varphi \\ u^2 &= \sinh\alpha \sin\beta \sin\varphi \\ u^3 &= \sinh\alpha \cos\beta. \end{aligned} \tag{18}$$

By substituting spin connection coefficients, basis, and 4-velocity to the equation (17) then we get

$$\begin{aligned} F_g^1 &= -\Omega^1_{\mu B}(x) e^{\mu}_C(x) u^B u^C \\ &= -\nu' e^{-\lambda} \cosh^2\alpha + \frac{e^{-\lambda}}{r} \sinh^2\alpha \sin^2\beta \sin^2\varphi + \frac{e^{-\lambda}}{r} \sinh^2\alpha \cos^2\beta \\ F_g^2 &= -\Omega^2_{\mu B}(x) e^{\mu}_C(x) u^B u^C \\ &= -\frac{e^{-\lambda}}{r} \sinh^2\alpha \sin^2\beta \sin\varphi \cos\varphi + \frac{\cot\theta}{r} \sinh^2\alpha \cos^2\beta \\ F_g^3 &= -\Omega^3_{\mu B}(x) e^{\mu}_C(x) u^B u^C \\ &= -\frac{e^{-\lambda}}{r} \sinh^2\alpha \sin\beta \cos\beta \cos\varphi - \frac{\cot\theta}{r} \sinh^2\alpha \sin\beta \cos\beta \sin\varphi \end{aligned} \tag{19}$$

In the hyperbolic coordinate system the equation (19) are given by

$$\begin{aligned} F_g^{\alpha} &= (\cosh\alpha)^{-1}[\sin\beta(\cos\varphi F_g^1 + \sin\varphi F_g^2) + \cos\beta F_g^3] \\ &= -\nu' e^{-\lambda} \cosh\alpha \sin\beta \cos\varphi \\ F_g^{\beta} &= (\sinh\alpha)^{-1}[\cos\beta(\cos\varphi F^1 + \sin\varphi F^2) - \sin\beta F^3] \\ &= -\nu' e^{-\lambda} \coth\alpha \cosh\alpha \cos\beta \cos\varphi + \frac{e^{-\lambda}}{r} \sinh\alpha \cos\beta \cos\varphi + \frac{\cot\theta}{r} \sinh\alpha \cos\beta \sin\varphi \\ F_g^{\varphi} &= (\sinh\alpha)^{-1}(\sin\beta)^{-1}[-\sin\varphi F_g^1 + \cos\varphi F_g^2] \\ &= \frac{1}{\sinh\alpha \sin\beta} \left\{ \nu' e^{-\lambda} \cosh^2\alpha \sin\varphi - \frac{e^{-\lambda}}{r} \sinh^2\alpha \sin\varphi + \frac{\cot\theta}{r} \sinh^2\alpha \cos^2\beta \cos\varphi \right\}. \end{aligned} \tag{20}$$

The random impact of surrounding particles generally cause two kind of effects, that are, they act as a random driving force leading to a random motion and they give rise to a frictional force. In the non-relativistic theory the friction force is given by $F_f^i = -\kappa m u^i$, where κ is the friction coefficient and u^i are the components of the non-relativistic velocity. The relativistic generalization of the friction force requires the introduction of a friction tensor κ^i_{α} similar to the pressure tensor in the relativity theory. The friction force is expressed as [5], [8]

$$F_f^i = \kappa^i_{\alpha}[u^{\alpha} - V^{\alpha}] \tag{21}$$

where V^α is the 4-velocity of the heat bath and κ^i_α is friction coefficient tensor. For an isotropic homogeneous heat bath the friction tensor is given by

$$\kappa^i_\alpha = \kappa(\eta^i_\alpha + u^i u_\alpha). \tag{22}$$

where κ denoting the scalar friction coefficient measured in the rest frame of the particles. In the laboratory frame the heat bath is at rest described by $U^\alpha = (1,0,0,0)$. Therefore the friction force is given by $F^i_f = -\kappa u^i u^0$ or in hyperbolic coordinates

$$F^\alpha_f = -\kappa \sinh\alpha; F^\beta_f = 0; F^\varphi_f = 0, \tag{23}$$

The components of total force of a system in hyperbolic coordinate are given by

$$\begin{aligned}
F^\alpha &= F^\alpha_g + F^\alpha_f \\
&= -\nu' e^{-\lambda}\cosh\alpha\sin\beta\cos\varphi - \nu\sinh\alpha \\
F^\beta &= F^\beta_g + F^\beta_f \\
&= -\nu' e^{-\lambda}\coth\alpha\cosh\alpha\cos\beta\cos\varphi + \frac{e^{-\lambda}}{r}\sinh\alpha\cos\beta\cos\varphi + \frac{\cot\theta}{r}\sinh\alpha\cos\beta\sin\varphi \\
F^\varphi &= F^\varphi_g + F^\varphi_f \\
&= \frac{1}{\sinh\alpha\sin\beta}\left\{\nu' e^{-\lambda}\cosh^2\alpha\sin\varphi - \frac{e^{-\lambda}}{r}\sinh^2\alpha\sin\varphi + \frac{\cot\theta}{r}\sinh^2\alpha\cos^2\beta\cos\varphi\right\}
\end{aligned} \tag{24}$$

Diffusion equation that describes the diffusion of particles around a neutron star can be derived by calculating the divergence in the position space, the divergence in the velocity space, and Laplace-Beltrami operator. The divergence in the velocity space in the hyperbolic coordinate system are given by

$$\begin{aligned}
div_v(F\Phi) &= (\sinh\alpha)^{-2}\frac{\partial}{\partial\alpha}((\sinh\alpha)^2 F^\alpha\Phi) - (\sinh\alpha)^{-1}(\sin\beta)^{-1}\frac{\partial}{\partial\beta}(\sin\beta F^\beta\Phi) \\
&\quad -(\sinh\alpha)^{-1}(\sin\beta)^{-1}\frac{\partial}{\partial\varphi}(F^\varphi\Phi) \\
div_v(F\Phi) &= -2\nu'\Phi e^{-\lambda}\coth\alpha\cosh\alpha\sin\beta\cos\varphi - \nu'\Phi e^{-\lambda}\sinh\alpha\sin\beta\cos\varphi \\
&\quad -\nu' e^{-\lambda}\frac{\partial\Phi}{\partial\alpha}\cosh\alpha\sin\beta\cos\varphi - 3\nu\Phi\cosh\alpha - \nu\frac{\partial\Phi}{\partial\alpha}\sinh\alpha \\
&\quad +\nu'\Phi e^{-\lambda}\coth^2\alpha\csc\beta\cos\varphi - \frac{e^{-\lambda}\Phi}{r}\csc\beta\cos\varphi \\
&\quad -\frac{\Phi}{r}\cot\theta\csc\beta\sin\varphi - 2\nu'\Phi e^{-\lambda}\coth^2\alpha\sin\beta\cos\varphi + \frac{2\Phi e^{-\lambda}}{r}\sin\beta\cos\varphi \\
&\quad +\frac{2\Phi}{r}\cot\theta\sin\beta\sin\varphi + \nu' e^{-\lambda}\frac{\partial\Phi}{\partial\beta}\coth^2\alpha\cos\beta\cos\varphi \\
&\quad -\frac{e^{-\lambda}}{r}\frac{\partial\Phi}{\partial\beta}\cos\beta\cos\varphi - \frac{1}{r}\frac{\partial\Phi}{\partial\beta}\cot\theta\cos\beta\sin\varphi \\
&\quad -\nu'\Phi e^{-\lambda}\coth^2\alpha\csc^2\beta\cos\varphi - \nu' e^{-\lambda}\frac{\partial\Phi}{\partial\varphi}\coth^2\alpha\csc^2\beta\sin\varphi \\
&\quad +\frac{e^{-\lambda}\Phi}{r}\csc^2\beta\cos\varphi + \frac{e^{-\lambda}}{r}\frac{\partial\Phi}{\partial\varphi}\csc^2\beta\sin\varphi \\
&\quad -\frac{1}{r}\frac{\partial\Phi}{\partial\varphi}\cot\theta\cot^2\beta\cos\varphi + \frac{\Phi}{r}\cot\theta\cot^2\beta\sin\varphi.
\end{aligned} \tag{25}$$

The divergence in the position space $div_x(e_M(x)\Phi)$ in the hyperbolic coordinate system are given by

$$\begin{aligned}
-u^M div_x(e_M(x)\Phi) &= -u^M \frac{1}{\sqrt{g}}\frac{\partial}{\partial x^i}(\sqrt{g}e^i_M(x)\Phi) \\
&= -\nu'\Phi e^{-\lambda}\sinh\alpha\sin\beta\cos\varphi - \frac{2e^{-\lambda}\Phi}{r}\sinh\alpha\sin\beta\cos\varphi \\
&\quad -e^{-\lambda}\frac{\partial\Phi}{\partial r}\sinh\alpha\sin\beta\cos\varphi - \frac{\Phi}{r}\cot\theta\sinh\alpha\sin\beta\sin\varphi \\
&\quad -\frac{1}{r}\frac{\partial\Phi}{\partial\theta}\sinh\alpha\sin\beta\sin\varphi - \frac{1}{r}\frac{\partial\Phi}{\partial\phi}\csc\theta\sinh\alpha\cos\beta
\end{aligned}$$

Laplace-Beltrami operator can be determined by using below equation

$$\begin{aligned}
\frac{D}{2}\Delta_u\Phi &= \frac{D}{2}\left\{\frac{\partial^2\Phi}{\partial\alpha^2} + 2\coth\alpha\frac{\partial\Phi}{\partial\alpha} - \frac{1}{(\sinh\alpha)^2}\left(\frac{\partial^2\Phi}{\partial\beta^2} + \cot\beta\frac{\partial\Phi}{\partial\beta}\right.\right. \\
&\quad \left.\left. + \frac{1}{(\sin\beta)^2}\frac{\partial^2\Phi}{\partial\varphi^2}\right)\right\} \\
\frac{D}{2}\Delta_u\Phi &= \frac{D}{2}\left\{\frac{\partial^2\Phi}{\partial\alpha^2} + 2\coth\alpha\frac{\partial\Phi}{\partial\alpha} + \frac{J(J+1)\Phi}{(\sinh\alpha)^2}\right\}, \qquad (26)
\end{aligned}$$

where $-J(J+1) = \frac{\partial^2}{\partial\beta^2} + \cot\beta\frac{\partial}{\partial\beta} + \frac{1}{(\sin\beta)^2}\frac{\partial^2}{\partial\varphi^2}$ and J is the discrete index with the values $J = 0, 1, 2, \ldots$.

The general relativistic diffusion equation of difusive particles around non-rotating neutron star in the parametrization of observer time in the hyperbolic coordinate system are given by

$$\begin{aligned}
e^{-\nu}\cosh\alpha\frac{\partial\Phi}{\partial x^0} &= -\frac{2e^{-\lambda}\Phi}{r}\sinh\alpha\sin\beta\cos\varphi - e^{-\lambda}\frac{\partial\Phi}{\partial r}\sinh\alpha\sin\beta\cos\varphi \\
&\quad -\frac{\Phi}{r}\cot\theta\sinh\alpha\sin\beta\sin\varphi - \frac{1}{r}\frac{\partial\Phi}{\partial\theta}\sinh\alpha\sin\beta\sin\varphi \\
&\quad -\frac{1}{r}\frac{\partial\Phi}{\partial\phi}\csc\theta\sinh\alpha\cos\beta + 2\nu'\Phi e^{-\lambda}\coth\alpha\cosh\alpha\sin\beta\cos\varphi \\
&\quad +\nu' e^{-\lambda}\frac{\partial\Phi}{\partial\alpha}\cosh\alpha\sin\beta\cos\varphi + 3\nu\Phi\cosh\alpha + \nu\frac{\partial\Phi}{\partial\alpha}\sinh\alpha \\
&\quad -\nu'\Phi e^{-\lambda}\coth^2\alpha\csc\beta\cos\varphi + 2\nu'\Phi e^{-\lambda}\coth^2\alpha\sin\beta\cos\varphi \\
&\quad +\frac{\Phi}{r}\cot\theta\csc\beta\sin\varphi + \frac{e^{-\lambda}\Phi}{r}\csc\beta\cos\varphi - \frac{2\Phi e^{-\lambda}}{r}\sin\beta\cos\varphi \\
&\quad -\frac{2\Phi}{r}\cot\theta\sin\beta\sin\varphi - \nu' e^{-\lambda}\frac{\partial\Phi}{\partial\beta}\coth^2\alpha\cos\beta\cos\varphi \\
&\quad +\frac{e^{-\lambda}}{r}\frac{\partial\Phi}{\partial\beta}\cos\beta\cos\varphi + \frac{1}{r}\frac{\partial\Phi}{\partial\beta}\cot\theta\cos\beta\sin\varphi \\
&\quad +\nu'\Phi e^{-\lambda}\coth^2\alpha\csc^2\beta\cos\varphi - \frac{e^{-\lambda}\Phi}{r}\csc^2\beta\cos\varphi \\
&\quad +\nu' e^{-\lambda}\frac{\partial\Phi}{\partial\varphi}\coth^2\alpha\csc^2\beta\sin\varphi - \frac{e^{-\lambda}}{r}\frac{\partial\Phi}{\partial\varphi}\csc^2\beta\sin\varphi \\
&\quad +\frac{1}{r}\frac{\partial\Phi}{\partial\varphi}\cot\theta\cot^2\beta\cos\varphi - \frac{\Phi}{r}\cot\theta\cot^2\beta\sin\varphi + \\
&\quad +\frac{D}{2}\left\{\frac{\partial^2\Phi}{\partial\alpha^2} + 2\coth\alpha\frac{\partial\Phi}{\partial\alpha} + \frac{J(J+1)\Phi}{(\sinh\alpha)^2}\right\}. \qquad (27)
\end{aligned}$$

5. The Physical Meaning

In contrast to Herrmann's work, it's difficult to determine general solution of our equation (27). However, some of physical meanings of the equation can be concluded qualitatively. Equation (27) can be written as

$$\begin{aligned}
&\frac{\partial\Phi}{\partial x^0} + e^{\nu-\lambda}\frac{\partial\Phi}{\partial r}\tanh\alpha\sin\beta\cos\varphi + \frac{e^{\nu}}{r}\frac{\partial\Phi}{\partial\theta}\tanh\alpha\sin\beta\sin\varphi + \frac{e^{\nu}}{r}\frac{\partial\Phi}{\partial\phi}\csc\theta\tanh\alpha\cos\beta = \\
&-\frac{2e^{\nu-\lambda}\Phi}{r}\tanh\alpha\sin\beta\cos\varphi - \frac{e^{\nu}\Phi}{r}\cot\theta\tanh\alpha\sin\beta\sin\varphi + 2\nu'\Phi e^{\nu-\lambda}\coth\alpha\sin\beta\cos\varphi \\
&+\nu' e^{\nu-\lambda}\frac{\partial\Phi}{\partial\alpha}\sin\beta\cos\varphi + 3\nu e^{\nu}\Phi + \nu e^{\nu}\frac{\partial\Phi}{\partial\alpha}\tanh\alpha - \nu'\Phi e^{\nu-\lambda}\coth\alpha\,\mathrm{csch}\,\alpha\csc\beta\cos\varphi \\
&+2\nu'\Phi e^{\nu-\lambda}\coth\alpha\,\mathrm{csch}\,\alpha\sin\beta\cos\varphi + \frac{e^{\nu}\Phi}{r}\cot\theta\,\mathrm{sech}\,\alpha\csc\beta\sin\varphi + \frac{e^{\nu-\lambda}\Phi}{r}\mathrm{sech}\,\alpha\csc\beta\cos\varphi \\
&-\frac{2\Phi e^{\nu-\lambda}}{r}\mathrm{sech}\,\alpha\sin\beta\cos\varphi - \frac{2e^{\nu}\Phi}{r}\cot\theta\,\mathrm{sech}\,\alpha\sin\beta\sin\varphi - \nu' e^{\nu-\lambda}\frac{\partial\Phi}{\partial\beta}\coth\alpha\,\mathrm{csch}\,\alpha\cos\beta\cos\varphi \\
&+\frac{e^{\nu-\lambda}}{r}\frac{\partial\Phi}{\partial\beta}\mathrm{sech}\,\alpha\cos\beta\cos\varphi + \frac{e^{\nu}}{r}\frac{\partial\Phi}{\partial\beta}\cot\theta\,\mathrm{sech}\,\alpha\cos\beta\sin\varphi + \nu'\Phi e^{\nu-\lambda}\coth\alpha\,\mathrm{csch}\,\alpha\csc^2\beta\cos\varphi \\
&-\frac{e^{\nu-\lambda}\Phi}{r}\mathrm{sech}\,\alpha\csc^2\beta\cos\varphi + \nu' e^{\nu-\lambda}\frac{\partial\Phi}{\partial\varphi}\coth\alpha\,\mathrm{csch}\,\alpha\csc^2\beta\sin\varphi - \frac{e^{\nu-\lambda}}{r}\frac{\partial\Phi}{\partial\varphi}\mathrm{sech}\,\alpha\csc^2\beta\sin\varphi \\
&+\frac{e^{\nu}}{r}\frac{\partial\Phi}{\partial\varphi}\cot\theta\,\mathrm{sech}\,\alpha\cot^2\beta\cos\varphi - \frac{e^{\nu}\Phi}{r}\cot\theta\,\mathrm{sech}\,\alpha\cot^2\beta\sin\varphi + \\
&+\frac{e^{\nu}D}{2}\left\{\mathrm{sech}\,\alpha\frac{\partial^2\Phi}{\partial\alpha^2} + 2\,\mathrm{csch}\,\alpha\frac{\partial\Phi}{\partial\alpha} + \frac{\mathrm{sech}\,\alpha}{(\sinh\alpha)^2}J(J+1)\Phi\right\}. \qquad (28)
\end{aligned}$$

Let compare the equation (28) and (5). The first term in the left hand side of equation (28) is the derivative of probability density function with respect to time coordinate, whereas the others term are the derivative of probability density function with respect to space coordinates. The last term in the right hand side of equation (28) is Laplace-Beltrami operator in the hyperbolic coordinate system where D is diffusion constant. The others term in the right hand side are predicted as the source of diffusion due to the curvature of spacetime which depend on 4-velocity.

Let consider the equation in the asymptotic case for which $r \to \infty$ and $e^{\nu} = e^{\lambda} = 1$. In this case, we obtain

$$\begin{aligned}
\frac{\partial\Phi}{\partial x^0} + \frac{\partial\Phi}{\partial r}\tanh\alpha\sin\beta\cos\varphi \;=\;& \left(\nu'\sin\beta\cos\varphi + \nu\tanh\alpha\right)\frac{\partial\Phi}{\partial\alpha} - \left(\nu'\coth\alpha\,\mathrm{csch}\,\alpha\cos\beta\cos\varphi\right)\frac{\partial\Phi}{\partial\beta} \\
&+\left(\nu'\coth\alpha\,\mathrm{csch}\,\alpha\csc^2\beta\sin\varphi\right)\frac{\partial\Phi}{\partial\varphi} + 2\nu'\Phi\coth\alpha\sin\beta\cos\varphi \\
&+3\nu\Phi + 2\nu'\Phi\coth\alpha\,\mathrm{csch}\,\alpha\sin\beta\cos\varphi \\
&-\nu'\Phi\coth\alpha\,\mathrm{csch}\,\alpha\csc\beta\cos\varphi + \nu'\Phi\coth\alpha\,\mathrm{csch}\,\alpha\csc^2\beta\cos\varphi \\
&+\frac{D}{2}\left\{\mathrm{sech}\,\alpha\frac{\partial^2\Phi}{\partial\alpha^2} + 2\,\mathrm{csch}\,\alpha\frac{\partial\Phi}{\partial\alpha} + \frac{\mathrm{sech}\,\alpha}{(\sinh\alpha)^2}J(J+1)\Phi\right\}. \qquad (29)
\end{aligned}$$

In the asymptotic case, the diffusion therefore depends only on the radial coordinate as well as the 4-velocity.

6. Conclusion

The diffusion equation of diffusive particles around a non-rotating neutron star is given by (27). The additional terms on the right hand side of equation (28) are predicted as the source of

Journal of Physics: Conference Series **539** (2014) 012005 doi:10.1088/1742-6596/539/1/012005

diffusion due to the curvature of spacetime which depend on 4-velocity. In the asymptotic case, the diffusion depends only on the radial coordinate as well as the 4-velocity.

7. References

[1] Anderson J L 1967 *Principles of Relativity Physics* (New York: Academic Press)
[2] Beskin V S 2010 *MHD Flows in Compact Astrophysical Objects: Accretion, Winds and Jets* (Verlag: Springer)
[3] Camenzind M 2007 *Compact Objects In Astrophysics: White Dwarfs, Neutron Stars, and Black Holes*, (Berlin: Springer)
[4] Dunkel J and Hanggi P 2005 *Phys. Rev. E* **71** 016124
[5] Dunkel J and Hanggi P 2005 *Phys. Rev. E* **72** 036106
[6] Haensel P, Potekhin A Y and Yakovlev D G 2007 *Neutron Stars 1, Equation of State and Structure* (New York: Springer)
[7] Herrmann J 2009 *Phys. Rev. E* **80** 051110
[8] Herrmann J 2010 *Phys. Rev. E* **82** 024026
[9] Ikeda N and Watanabe S 1989 *Stochastic Differential Equations and Diffusion Processes*, (Amsterdam: North-Holland Publishing Company)
[10] Kao E P C 1997 *An Introduction to Stochastic Process* (Houston: Duxbury Press)
[11] Morters P and Peres Y 2010 *Brownian Motion* (Cambridge: Cambridge University Press)
[12] Van Kampen N G 2007 *Stochastic Processes in Physics and Chemistry* (Amsterdam: North-Holland Publishing Company)

Magnetic Reconnection of Solar Flare Detected by Solar Radio Burst Type III

Z.S. Hamidi[1], N.N.M.Shariff, Z.A. Ibrahim, C. Monstein, W.N.A. Wan Zulkifli, M.B. Ibrahim, N.S. Arifin, N. A. Amran
School of Physics and Material Sciences, Faculty of Applied Sciences, MARA University of Technology, 40450, Shah Alam, Selangor, Malaysia
Academy of Contemporary Islamic Studies (ACIS), MARA University of Technology, 40450, Shah Alam, Selangor, Malaysia
Department of Physics, Faculty of Science, University of Malaya, 50603 Kuala Lumpur, Malaysia
Institute of Astronomy, Wolfgang-Pauli-Strasse 27, Building HIT, Floor J, CH-8093 Zurich, Switzerland

E-mail: zetysh@salam.uitm.edu.my

Abstract. The Sun is an ideal object of a blackbody with a large and complex magnetic field. In solar activity specifically solar flare phenomenon, the magnetic reconnection is one of the most significant factors of the Sun that can simplify a better understanding of our nearest star. This factor is due to the motion of the plasma and other particles through the convection mechanism inside the Sun. In our work, we will highlight one of the solar burst events that associated with solar flares. This event occurred on 13th November 2012 from 2:00:03 UT till 2:00:06 UT. It peaked with M2.0 solar flare at 2.04 UT. Within short time intervals of about $10^2 \sim 10^3$s, large quantities of energy of $10^{22} \sim 10^{26}$J are emancipated. The changing magnetic field converts magnetic potential energy into kinetic energy by accelerating plasmas in the solar corona. It is believed that the plasma is channelled by the magnetic field up and away from the Sun. It is also accelerated back down along the magnetic field into the chromosphere. In conclusion, we showed that the structure of the solar radio burst type III is an indicator of a starting point of magnetic reconnection.

1. Introduction

The Sun, a main sequence star with spectral type of G2V is considered as one of the strongest radio sources and observation in this widely region can provide evidence on structures throughout the solar atmosphere. It is widely known that the transportation of energy from the central regions of the Sun is primarily through photon radiation, although electron conduction contributes in the innermost region and convection dominates near the surface. However, the dynamical behaviour of the Sun exhibits a variety of physical phenomena, some of which are still not at all or only barely understood due to the complexity of the structure of the Sun. Along the process, many research works have done to fulfil the objectives of this study.

[1] School of Physics and Material Sciences, Faculty of Applied Sciences, MARA University of Technology, 40450, Shah Alam, Selangor, Malaysia

Meanwhile, the corona extends from the top of a narrow transition region to Earth and beyond has a temperature millions of degrees which is still considered as a mysteries properties. It is composed by protons, electrons an uncertain amount of helium nuclei and a small portion of highly ionized heavier atoms. It can be observed clearly during eclipse and where the solar wind originates. A multi-wavelength observation is believed a good approach in order to understand the variability of solar wind [1].

Before going into details of observations, it would be instructive to give some basic information on solar activities such as solar flares and Coronal Mass Ejections (CMEs). These phenomena can stimulate plasma oscillations which can release radiation at metric and decametric wavelengths. Solar flare is considered as a high energetic and complicated phenomenon in which mass eruptions occur, energetic particles are generated and highly energy radiations are emitted. In general, there are 3 types of acceleration commonly quoted in solar flares which is (i) Direct electric field acceleration [2,3] can boost a particle to high energies simply via the Coulomb force from the electric field and may drive in the current sheet or in the reconnection site, but it is hard to maintain a large-scale coherent direct current, DC electric field. (ii) Shock or first-order Fermi acceleration can energize particles by making them repeatedly pass through the shock front back and forth and this mechanism may be present in the fast shock produced by the super-magnetosonic outflow jet from the reconnection region. However, it would be difficult to reflect the particles in the upstream region. (iii) Stochastic (second- order Fermi) acceleration by turbulence or plasma waves is the most likely mechanism [4,5] for solar flares, compared with the shortcomings of the other 2 mechanisms.

During a flares, large quantities of energy are transferred between the corona and chromosphere through thermal conduction, non-thermal particle beams, radiation transport and the mass motions [6]. This event is triggered by fast drift of the individual active region proper motion within the complex magnetic configuration due to instabilities of equilibrium of coronal magnetic field [7]. Since the magnetic force is much stronger than other forces in the corona, any coronal structure is mainly controlled by magnetic field.

Meanwhile, metric radio burst is normally a non-thermal particles accelerated and trapped during those events [6] . In the presence of flares, strong solar radio burst type III is observed over a broad frequency range. It is generally believed that the conversion of Langmuir wave to electromagnetic wave is the generating mechanism of solar type III bursts [8]. The Langmuir wave can be easily excited by the instability of the electron-beam current, and the interaction of the forward and backward Langmuir waves are considered to be the cause for the second harmonic of a type III burst, however, the generating mechanism of the backward Langmuir wave is not easy to be understood. Whereas it is well established that flares do not drive CMEs, both of them correspond to different aspects of the same magnetic energy release [6].

In the currently widely accepted scenario, the basic physical processes involved and the observational signatures are as follows. Magnetic reconnection, as the primary energy release mechanism occurring high in the corona, rapidly heats the plasma and accelerates particles. It is well known that process by which magnetic lines of force break and re-join into a lower-energy configuration, which magnetic energy is converted into plasma kinetic energy. The impulsive phase of the flare coincides with the acceleration phase of the CMEs Some particles escape along the open magnetic field lines into the interplanetary space, with electrons producing various radio bursts and some electrons [9] and ions being detected at 1 AU. In what manner and when the magnetic reconnection in the solar flare occur is remain as one of the longest outstanding questions in solar physics, holding clues about the onset of structure formation in the first light of solar burst detection. It is believed that metric radio observation plays an important role in understanding the behaviour of magnetic reconnection due to solar burst type III characteristic [10,11].

Since the corona is a good conductive medium, still it has a finite resistivity[12]. The magnetic energy can be released in the diffusion process. The region where the energy release works efficiently is known as a current sheet, in which a certain component of the magnetic field is abruptly changed across the sheet. This energy is thought to be released high in the corona and produces emission across the entire electromagnetic spectrum, from radio to gamma ray energies and heats the surrounding plasma to tens of Kelvins. Roughly, around 50% of the energy released in a flare goes into accelerating particles [13,14]. It is thought that the energy required for a flare is stored as 'free' magnetic energy in the corona in the form of a current .

2. Experimental Setup and Observations

The *Compound Low Cost Low Frequency Transportable Observatory* (CALLISTO) system has been operating routinely at the National Space Centre, Sg Lang Selangor (3.0833333°N 101.5333333°E) since March 2012 funded by the National Space Centre, Malaysia[15,16]. This is one of the radio quiet zone candidate sites in Malaysia. It obtains the minimum Radio Frequency Interference (RFI) with average -102 dBm [17]. E-Callisto networking is an internationalization research collaboration that investigate the phenomena of solar flares and Coronal Mass Ejections based on solar burst characteristics in a broadband radio region from 45 -870 MHz [18]. This collaboration has achieved the objective to monitor the Sun within 24 hours all over the world; the precise data will lead to an understanding of the causes of spectacular ejection of material from flare region that associated with the coronal magnetic field [19]. So far, we have 2 spectrometers that will be installed at INSTUN, Perak and Jelebu, Negeri Sembilan. Figure 1 shows the schematic diagram of CALLISTO system [20].

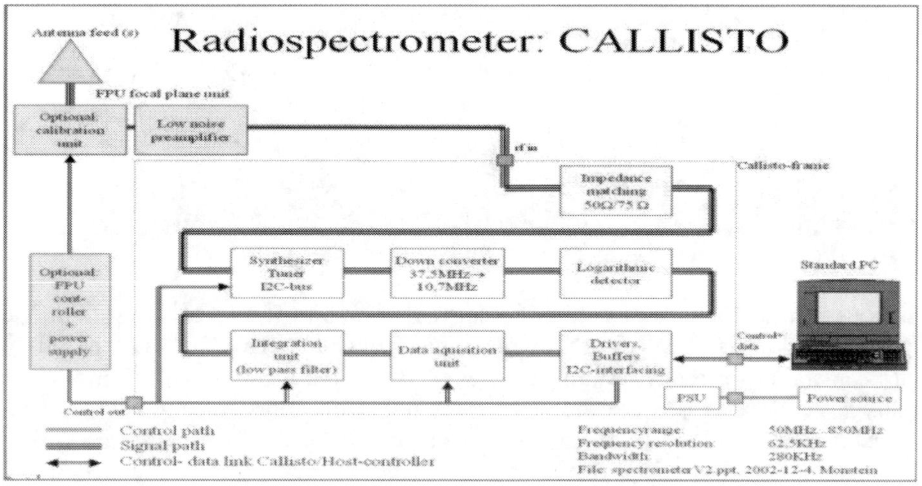

Figure 1. Schematic diagram CALLISTO system

3. Results and Discussion

In this section, a comprehensive analysis of solar radio burst Type III observation of short-duration in association with eruptions on the sun and appear like a radio burst Type II is presented. This observation allows for the mechanisms of evolution Type II and local environment of the burst to be characterized. Yet, the burst is quite unique and interesting to be tackled. The burst is classified as a solar radio burst type III. However, a clear structure of fundamental solar radio burst Type II can be observed.

This event occurred on 13th November 2012 from 2:00:03 UT till 2:00:06 UT. It peaked with M2.0 solar flare at 2.04 UT. From the earliest prediction, there is a possibility a Coronal Mass Ejections (CMEs) will eject directly to the Earth. This is due to a continuous monitoring within few days period. Based on the results, it is associated with type M 6.0 solar flares from active region AR1613 and AR1614. It was found that the sunspot number is 188 and radio flux exceeds 144 sfu. There are 8 active regions and with a density of 9.1 protons/ cm^3. During that event, solar wind also reached up to 369.7 km sec^{-1}. Those parameters provided an alert with Sun's activities. The potential of a big explosion of CMEs is considered very high. Based on observation, the formation of solar radio burst Type III is due to multiple radio blackout storms.

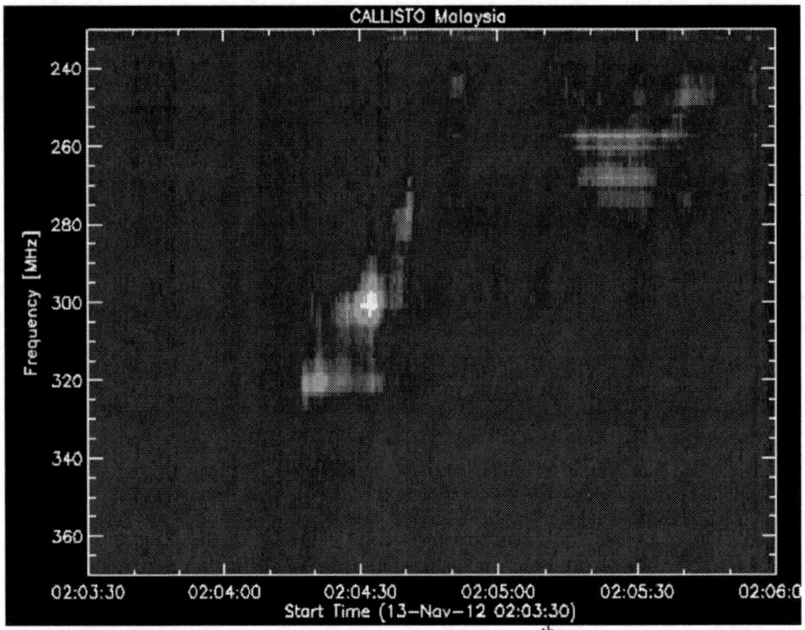

Figure 2. Solar Radio Burst Type III on 13th November 2012

As can be seen from Figure 2, there is a group of solar radio burst Type III (at the background of the intense burst) that formed intermittently with a wide range frequency from 220 MHz till 380 MHz. In conjunction with constraints on the spectrum, it is believed that the burst is drifting in a wide range. This burst actually has been observed intermittently in the last a few hours since 00:22 UT. The groups of solar radio burst Type III until 5:30 UT can still be observed. This proved that a meter solar radio burst will take a long period to decay and AR1613 and AR1614 potentially active within a few day comparison, the dynamic spectra in the X-ray region beginning from 1:00 UT till 7:00 UT is shown in Figure 3.

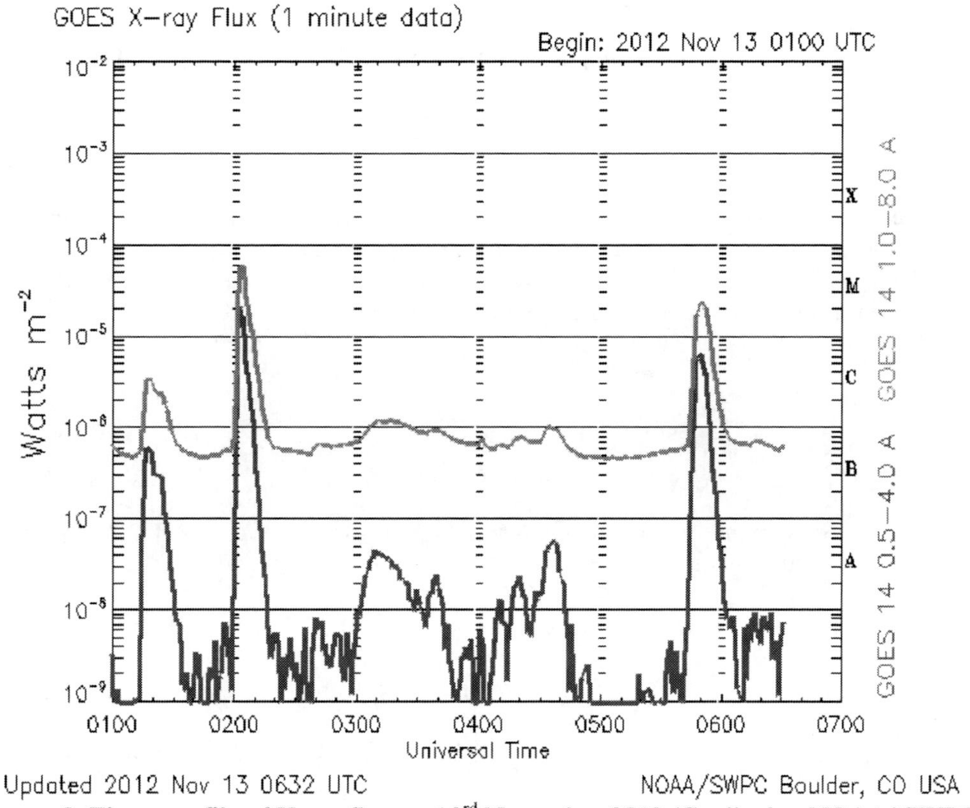

Figure 3. Times profile of X-ray flux on 13rd November 2012 (Credited to NOAA/ SWPC)

In particular, an observed random variable may be represented as instability of the magnetic field of the active region. There was a CMEs event (~ 383 kms-1) that has been detected at 1:28 UT from LASCO data. As can be seen the X-ray flux is approximately to the 10-4 Watts m-2 during 2:00 UT. This is the highest point within six hours period of time. The sun remains active with 2 solar flares class M can be detected.

Overall, the Sun tends to be consistent with multiple storm, however, not huge CMEs explosion. Based on LASCO observation, the CMEs explosion with a speed of 419 kms^{-1} can be detected at 3:12 UT which means that almost after 1 hour and 10 minutes after the formation of this burst. In conclusion, this event is one fine example of tendencies solar radio burst Type III, which makes it possible to form solar radio burst Type II. The burst is form in between of 2 events of CMEs in at the same day.

4. Conclusion

It is undeniable that solar flare plays an important role in the Sun-Earth connection due to sudden changes of strong magnetic fields in the Sun's corona. Within short time intervals of about $10^2 \sim 10^3$s, large quantities of energy of $10^{22} \sim 10^{26}$J are emancipated. The changing magnetic field converts magnetic potential energy into kinetic energy by accelerating plasmas in the corona. The plasma is channelled by the magnetic field up and away from the Sun. It is also accelerated back down along the magnetic field into the chromosphere. In the chromosphere, the plasma crashes into denser gas and releases its kinetic energy into thermal energy, sound, and light energy The burst impulsive phase onset could occur at first lower frequencies, although there are also examples with delays in the opposite sense. It is hoped that we could possible to find a latest data that might answered the mechanism of type III burst based on solar flare events.

Still, it is not fully understood how the energy is released and converted into thermal heat and into non-thermal particle acceleration energy. It should be noted; however, there are many complex models which propose how this occurs, yet most models at some point invoke magnetic reconnection. There are also other models which do not feature magnetic reconnection as the main cause of energy release. One particular model suggests electrons are accelerated in the corona to energies of up to 10 - 100's of keV. These electrons then propagate down the newly connected field lines. Upon reaching the loop foot points the electrons encounter an increased plasma density where they undergo Coulomb collisions and emit the hard X-ray emission.

5. References

[1] W.-H. Ip, Global Solar Wind Interaction and Ionospheric Dynamics, Comets II, 2004, pp. 605-629.
[2] S.G. Benka, G.D. Holman, ApJ 435 (1994).
[3] G.D. Holman, M.R. & Kundu, ApJ 292 (1985).
[4] V. Petrosian, High-Energy Solar Phenomena, in: J.M. Ryan, W.T. Vestrand (Eds.), AIP Conf. Proc. 294, New York: AIP, 1994.
[5] V. Petrosian, High Energy Solar Physics, in: R. Ramaty, N. Mandzhavidze, X.-M. Hua (Eds.), Woodbury: AIP, 1996.
[6] N. Gopalswamy, M.R. Kundu, Thermal and nonthermal emissions during a coronal mass ejection, Solar Physics 143 (1993) 327-343.
[7] D.Herdiwijaya, S.Imelda, The Probability of Flare Occurrences Based on Sunspot Group and Magnetic Configurations, Jurnal Matematika dan Sains 11 (2006) 7.
[8] D.J. McLean, a.N.R. Labrum, Solar Radiophysics, Cambridge University Press, Cambridge, 1985.
[9] Y. Wang, Zhang, A Comparative Study between Eruptive X-Class Flares Associated with Coronal Mass Ejections and Confined X-Class Flares, Astrophys. J. 665 (2007) 1428-1438.
[10] G.A. Dulk, Type III solar radio bursts at long wavelengths, in: R. Stone, E. Weiler, M. Goldstein (Eds.), Geophys. Monogr., 2000.
[11] D.B.Melrose, On the theory of Type II and Type III Solar Radio Bursts, Australian J. Phys. 23 (1970) 885-903.
[12] H. Aurass, Coronal Physics from Radio and Space Observations, in: I.G. Trottet (Ed.), Lecture Notes in Physics, Springer, Berlin, 1997.
[13] D.B.Melrose, A Solar Flare Model Based on Magnetic Reconnection between Current-carrying Loops, ApJ 486 (1994).
[14] A.G. Emslie, H. Kucharek, B.R. Dennis, N. Gopalswamy, G.D. Holman, G.H. Share, A. Vourlidas, T.G. Forbes, P.T. Gallagher, G.M. Mason, T.R. Metcalf, R.A. Mewaldt, R.J. Murphy, R.A. Schwartz, T.H. Zurbuchen, Energy partition in two solar flare//CME events, Journal of Geophysical Research 109 (2004) 15.
[15] Z.S. Hamidi, N. Shariff, Z. Abidin, Z. Ibrahim, C. Monstein, Coverage of Solar Radio Spectrum in Malaysia and Spectral Overview of Radio Frequency Interference (RFI) by Using CALLISTO Spectrometer from 1MHz to 900 MHz, Middle-East Journal of Scientific Research 12 (2012) 6.
[16] Z.S. Hamidi, Z. Ibrahim, Z. Abidin, M. Maulud, N. Radzin, N. Hamzan, N. Anim, N. Shariff, Designing and Constructing Log Periodic Dipole Antenna to Monitor Solar Radio Burst: e-Callisto Space Weather, International Journal of Applied Physics and Mathematics 2 (2012) 3.
[17] Z.S.Hamidi, Z. Abidin, Z. Ibrahim, N. Shariff, Indication of radio frequency interference (RFI) sources for solar burst monitoring in Malaysia, ICPAP 2011, AIP Publisher, Indonesia, 2012, pp. 6.

[18] Benz A.O. , C. Monstein, H. Meyer, CALLISTO, A New Concept for Solar Radio Spectrometers, Kluwer Academic Publishers, 2004.
[19] A.O. Benz, C. Monstein, H. Meyer, P.K. Manoharan, R. Ramesh, A. Altyntsev, A. Lara, J. Paez, K.-S. Cho, A World-Wide Net of Solar Radio Spectrometers: e-CALLISTO, Earth Moon and Planets 104 (2009) 277 – 285.
[20] Z.S.Hamidi, N.M.Anim, N. N.S.Hakimi, N.Hamzan, A.Mokhtar, N.Syukri, S.Rohizat, I.Sukma, Z.A. Ibrahim, Z.Z.Abidin, N.N.M.Shariff, C.Monstein, Application of Log Periodic Dipole Antenna (LPDA) in Monitoring Solar Burst at Low Region Frequencies Region International Journal of Fundamental Physical Sciences 2 (2012) 4.

Acknowledgments
This project is an initiative of the International Space Weather Initiative (ISWI) project. Special thanks to the Institute of Astrophysics, ETH Zurich, Switzerland for initiating the CALLISTO network, the University of Malaya for the grant (PV071/2011B) and (UM-HIR H-21001-F000028), MARA University of Technology (UITM) and the National Space Agency (ANGKASA) for providing a site to set up this project.

Hierarchical Bayesian Spatio Temporal Model Comparison on the Earth Trapped Particle Forecast

Wayan Suparta[1], **Gusrizal**[1]

[1] Space Science Centre (ANGKASA), Institute of Climate Change, Universiti Kebangsaan Malaysia, 43600 Bangi, Selangor Darul Ehsan, Malaysia

E-mail: wayan@ukm.my

Abstract. We compared two hierarchical Bayesian spatio temporal (HBST) results, Gaussian process (GP) and autoregressive (AR) models, on the Earth trapped particle forecast. Two models were employed on the South Atlantic Anomaly (SAA) region. Electron of >30 keV (mep0e1) from National Oceanic and Atmospheric Administration (NOAA) 15-18 satellites data was chosen as the particle modeled. We used two weeks data to perform the model fitting on a 5°x5° grid of longitude and latitude, and 31 August 2007 was set as the date of forecast. Three statistical validations were performed on the data, i.e. the root mean square error (RMSE), mean absolute percentage error (MAPE) and bias (BIAS). The statistical analysis showed that GP model performed better than AR with the average of RMSE = 0.38 and 0.63, MAPE = 11.98 and 17.30, and BIAS = 0.32 and 0.24, for GP and AR, respectively. Visual validation on both models with the NOAA map's also confirmed the superior of the GP than the AR. The variance of log flux minimum = 0.09 and 1.09, log flux maximum = 1.15 and 1.35, and in successively represents GP and AR.

1. Introduction

Trapped particle has a major role for low Earth orbit (LEO) satellite anomaly. The risk is increased for the LEO Equatorial inclination satellites due to their frequent pass over the South Atlantic Anomaly (SAA) region. The SAA is an area with dense distribution of trapped particle, especially proton and low energy electron [1]. The high energy proton causes the degradation of satellite solar array whereas the low energy electron causes the satellite surface charging [2]. Overcoming with this problem, we develop a LEO trapped particle distribution forecasting system based on a hierarchical Bayesian spatio temporal (HBST) model. The HBST is a statistical model that works for spatio temporal data in a Bayesian framework [3].

In our previous work [4], we have employed a simple HBST model, i.e. Gaussian Process (GP) [5]. In this work, we attempt to employ HBST-Autoregressive (AR) model [6], a modified version of the GP. The difference between the two models is the AR considering the previous day parameter in their true value calculation (Equation 11). Detail of both models is explained in Section 2.

As in our previous work [4], we still use the data taken by National Oceanic and Atmospheric Administration (NOAA) satellite. However, in this work we attempt to employ multiple satellites data instead of only using NOAA 15, i.e. NOAA 15-18. We also improved our gridding method by shrinking the size of our grid from 10°x10° of longitude and latitude in our previous work to become

5°x5° of longitude and latitude as in Figure 1a. The application of 5°x5° grid size makes the SAA is divided in 12 layers at the latitude direction and randomly select four points of validation in each layer (as in Figure 1b). The detail about the gridding process is explained in our previous paper [4].

Since we start to employ multiple satellites data, then we reduce the data input, from 30 days data in our previous work to 14 days data in this work. This step is taken to optimize the time consumed while running computational process. We choose to employ the implementation of both models on the 30 keV of electron energy level (mep0e1). We choose mep0e1 due to the sensitivity of the medium energy proton and electron detector (MEPED) sensor installed in the NOAA satellites. The area of South Atlantic Anomaly (SAA) is still used in this work. We set 31 August 2007 as the target date of the forecasting. Kriging interpolation method still used to draw the distribution of forecasted particle flux value over the SAA region. Detail about the Kriging interpolation of our work can be found in Suparta et al. [7].

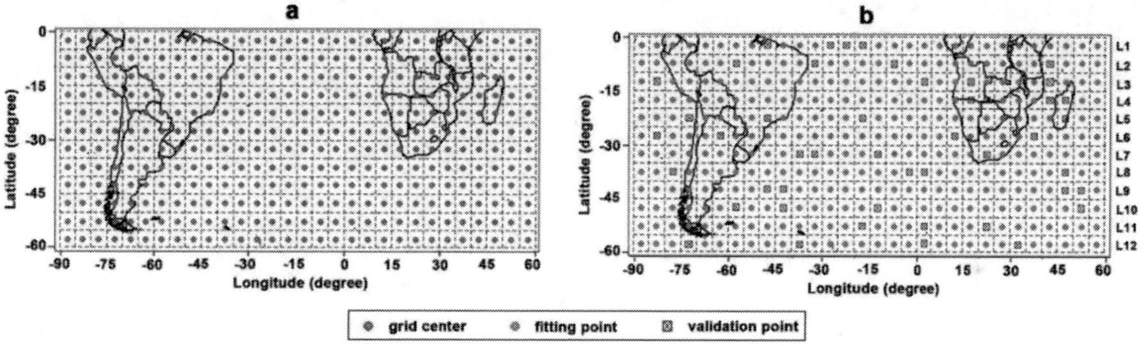

Figure 1. (a) A 5°x5° of longitude and latitude gridding system over SAA region, and (b) Fit and validation points over the SAA.

In the following section, we describe two HBST models mentioned, i.e. GP and AR models. Section 3 explains the forecasting results of the two models, performed in both statistical and visual analyses and discuss the results. The summary and future work is compiled in Section 4.

2. Hierarchical Bayesian Spatio Temporal Model

First, we define the notations used in this paper [8]. Let t defines time where $t=1,\ldots, T$, and T is the total number of time units. In this work, we put T = 14. This is done by selecting data from 17-30 August 2007 as the model input. For the forecasting day (T+1), we put 31 August 2007. The $Z(s_i, t)$ is a logarithmic value of SAA's 30 keV electron flux. We also denote $E(s_i, t)$ as a true value corresponding to $Z(s_i, t)$ at site s_i, $i = 1,\ldots,$ n at time t. Let $Z_t = (Z(s_i, t),\ldots, Z(s_n, t))'$ and $E_t = (E(s_i, t),\ldots, E(s_i, t))'$. We should denote all observed data by z, and z^* will yield all missing data. As for E, E defines all E_t, for t=1,…, T. The total number of observations modeled is defined by N, which $N=nT$. Generally, in Bayesian forecast modeling, there are p covariates, denote by X_t which are in the form of 1 x p vector. These covariates could be some variable that influence Z value. The regression coefficients are denoted by β, also in the form 1 x p vector. The NOAA data does not provide the covariates values for this work. Therefore, we only employ the intercept value rather than β. Finally, the θ symbol is used to denote all parameters used in our model ($\theta = \left(E, \beta, \sigma_\epsilon^2, \sigma_\eta^2, \emptyset\right)$).

2.1. Forecasting using Gaussian Process (GP) Models.

The spatio-temporal Bayesian GP model [5] is expressed as

$$Z_t = E_t + \epsilon_t \tag{1}$$

$$E_t = X_t\beta + \eta_t \tag{2}$$

where $\epsilon_t = \left(\epsilon(s_{1,}t),\dots,\epsilon(s_{n,}t)\right)' \sim N(0,\sigma_\epsilon^2 I_n)$ is a white noise error process. σ_ϵ^2 is the nugget effect, and is homogeneous in space and time. The term I_n is the *n* x *n* identity matrix. The term $\eta_t = \left(n(s_1,t),\dots,n(s_n,t)\right)' \sim N(0,\Sigma_n)$ is a spatially correlated error. $\Sigma_n = \sigma_\eta^2 S_n = \sigma_\eta^2\kappa(s_i,s_j;\phi,\upsilon)$ is a variance-covariance matrix with have dimension *n* x *n*, *i*, *j* = 1,…, *n*. σ_η^2 is the site invariant common variance, while $\kappa(.;\phi,\upsilon)$ is the spatial correlation matrix that have spatial decay ϕ and smoothness parameter υ. Both parameters, ϵ_t and η_t, are independent each other.
Hence, the GP model for 1-step ahead forecast for the distribution *Z* at any unobserved location S_0 is:

$$Z(s_0,T+1) = E(s_0,T+1) + \epsilon(s_0,T+1) \tag{3}$$
$$E(s_0,T+1) = x'(s_0,T+1)\beta + \eta(s_0,T+1) \tag{4}$$

the posterior predictive distribution of $Z(s_0,T+1)$ for a given **z** act as the 1-step ahead forecast distribution, i.e.

$$\pi(Z(s_0,T+1)|z) = \int \pi(Z(s_0,T+1)|\theta,E,E(s_0,T+1),z)\,\pi(E(s_0,T+1)|\theta,z)\pi(\theta,E|z)dE(s_0,T+1)dEd\theta \tag{5}$$

$\pi(\theta,E|z)$ is a joint posterior distribution of E and θ. Because of the conditional independence of $Z(s_0,T+1)$ for Z given E, so $\pi(Z(s_0,T+1)|\,\theta,E,E(s_0,T+1),z) = \pi\left(Z(s_0,T+1)|\,\theta,E,E(s_0,T+1)\right)$. As $E(s_0,T+1)$ does not depend on Z given E, then we can replace $\pi(E(s_0,T+1)|\theta,z)$ by $\pi(E(s_0,T+1)|\theta)$.

A 1-step ahead forecast distribution (5) can be constructed through a marginal distribution as well as conditional distribution. For marginal distribution, at the *j*th Monte Carlo Markov Chain (MCMC) iterations, we have posterior samples, $\theta^{(j)}$and $E^{(j)}$. $E^{(j)}(s_0,T+1)$ is drawn from $N\left(x'_{T+1}\beta^{(j)},\sigma_\eta^{2(j)}\right)$. $Z^{(j)}(s_0,T+1)$ is obtained from $N\left(E^{(j)}(s_0,T+1),\sigma_\eta^{2(j)}\right)$. To use conditional distribution, consider that the joint distribution of $E_{T+1} = \left(E(s_1,T+1),\dots,E(s_n,T+1)\right)'$ is given by $N\left(X_{T+1}\beta,\Sigma_\eta\right)$ and the conditional distribution of $\pi(E(s_0,T+1)|E_{T+1})$ is normally distributed with mean

$$x(s_0,T+1)\beta + S_{\eta,12}S_\eta^{-1}(E_{T+1} - X_{T+1}\beta) \tag{6}$$

and variance

$$\sigma_\eta^2(1 - S_{\eta,12}S_{\eta,21}^{-1}) \tag{7}$$

where $S'_{\eta,21} = S_{\eta,12} = e^{-\phi d_{12}}$ and $d_{12} = \left(||s_1 - s_0||,\dots,||s_n - s_0||\right)'$.

Similar with any unobserved point, a 1-step ahead forecast at any observed point is modeled by:

$$Z(s_i, T+1) = E(s_i, T+1) + \epsilon(s_i, T+1) \quad (8)$$

$$E(s_i, T+1) = x'(s_i, T+1)\beta + \eta(s_i, T+1) \quad (9)$$

the same step is applied for getting Z as in the unobserved point.

2.2. Forecasting using Auto-Regressive (AR) Models.

The AR model [6] is given by

$$Z_t = E_t + \epsilon_t \quad (10)$$

$$E_t = \rho E_{t-1} + X_t\beta + \eta_t \quad (11)$$

where ϵ_t and η_t form is defined as in the previous GP model. The ρ denotes the autoregressive process parameter, with $|\rho| < 1$. For $\rho = 0$, then our AR model turn to GP model as in (2). For predicting the distribution of $Z(s_0, T+1)$ at any unobserved point, it is determined by the value of $O(s_0, T+1)$. According to (11), $E(s_0, T+1)$ follows the normal distribution and has site variance σ_η^2 and mean $\rho E(s_0, T) + x'(s_0, T+1)$. Because dependent to $E(s_0, T)$, we have to specify all the random variables $E(s_0, T)$, for k=0,...,T. For the first step of simulation, we first simulate the conditional distribution of $E(s_0, 0)$ given E_0. Thus, at *j*th MCMC iterations, we consecutively simulate $E^{(j)}(s_0, k)$ given $E^{(j)}(s_0, k-1)$ for $k = 1, \ldots, T+1$ from a normal distribution with mean $\rho^{(j)} E^{(j)}(s_0, k-1) + x'(s_0, k)\beta^{(j)}$ and variance $\sigma_\eta^{2(j)}$.

Similarly, for forecasting at any observed point s_i, we draw $Z^{(j)}(s_i, T+1)$ from the normal distribution with mean $\rho^{(j)} E^{(j)}(s_i, T) + x'(s, T+1)\beta^{(j)}$ and variance $\sigma_\epsilon^{2(j)}$.

3. Results and Discussion

We use *R* ***spTimer*** package [9] to compute MCMC Gibbs sampling procedure as defined in previous section. We set 13,000 for *j* value and discharge the first 3,000 iterarations to avoid the starting effect. The forecasting acuracy of both models was then be evaluated by three statistical validations, i.e. the root mean square error (RMSE), mean absolute percentage error (MAPE) and bias (BIAS) on the validation points (Figure 1b).

The results of statistical validation of GP and AR models are displayed in Figure 2. Figures 2a and 2c, mentioned for RMSE and BIAS with units in log flux (particles/cm^2 sec str), whereas for MAPE in Figure 2b, the unit is in percentage. The accuracy of forecasting result is pointed by MAPE, which defines how close the forecasting result to the real data. Our models, GP and AR, give 11.98 and 17.30 for MAPE values which indicates their high degree of forecasting accuracy. On the other hand, RMSE result indicates variance between forecasting value and real data. We find that the average values of RMSE for GP and AR is 0.38 and 0.63, respectively. We could also evaluate the nature of flux distribution variability from the RMSE. The higher of RMSE value in a layer means the higher of variability among the flux in that layer. In this work, we could find that the L3 and L4 layers have the highest values of RMSE. It means that the high variability of flux particle distribution occurred at the SAA core. The trend of flux particle distribution variability shows that the distribution of flux is quite uniform outside the SAA area (L9 to L10). We also find that at SAA boundaries area such as L1, L2,

Journal of Physics: Conference Series **539** (2014) 012007 doi:10.1088/1742-6596/539/1/012007

L5-L8, the variability of flux starts to rise. In addition, we can accept the variability at L11 and L12 since there are boundary layers between the two dense areas, the SAA and the polar. From BIAS result, we find that our model mostly produced overestimate values in each layer except for L3, L7 and L8, where the AR gives the underestimate forecasts. The average values of BIAS for GP and AR are 0.32 and 0.24, respectively.

We perform a visual validation for the results as depicted in Figure 3. Figure 3a defines the plot of raw data acquired by NOAA 15-18 on 31 August 2007, whereas Figure 3b defines the NOAA map of mep0e1 on 31 August 2007. Figures 3c and 3d define the estimation of our forecast values after being interpolated by a Kriging technique, for GP and AR, respectively. The Kriging interpolation was computed by using *R* ***fields*** package [10]. The quality of Kriging interpolation is defined by its variance values and represented by Figures 3e and 3f, which successively represent GP and AR. The different between our results in Figures 3c-3d to the NOAA map in Figure 3b is the period of time. NOAA's map in Figure 3b is generated for 4 days distribution (28-31 August 2013) on a nowcast basis. On the other hand, our results in Figure 3c-3d showed forecast maps and designed for one day distribution, 31 August 2007.

As illustrated in Figures 3c-3d, the AR produced a significant overestimate number of a maximum and underestimate number of a minimum. The log maximum number of real data is 7.65 and 1.09 for a minimum. Thus, the GP gives 1.09 for a maximum variance and 0.09 for a minimum, whereas the AR gives 1.35 and 1.15 for a maximum and a minimum, respectively. From the comparison between Figures 3e and 3f, we could see that the GP result Kriging interpolation is better than the AR since it produced less variance than AR. From this result, we can conclude that the autoregressive parameter (in Equation 11) is not suitable for our work.

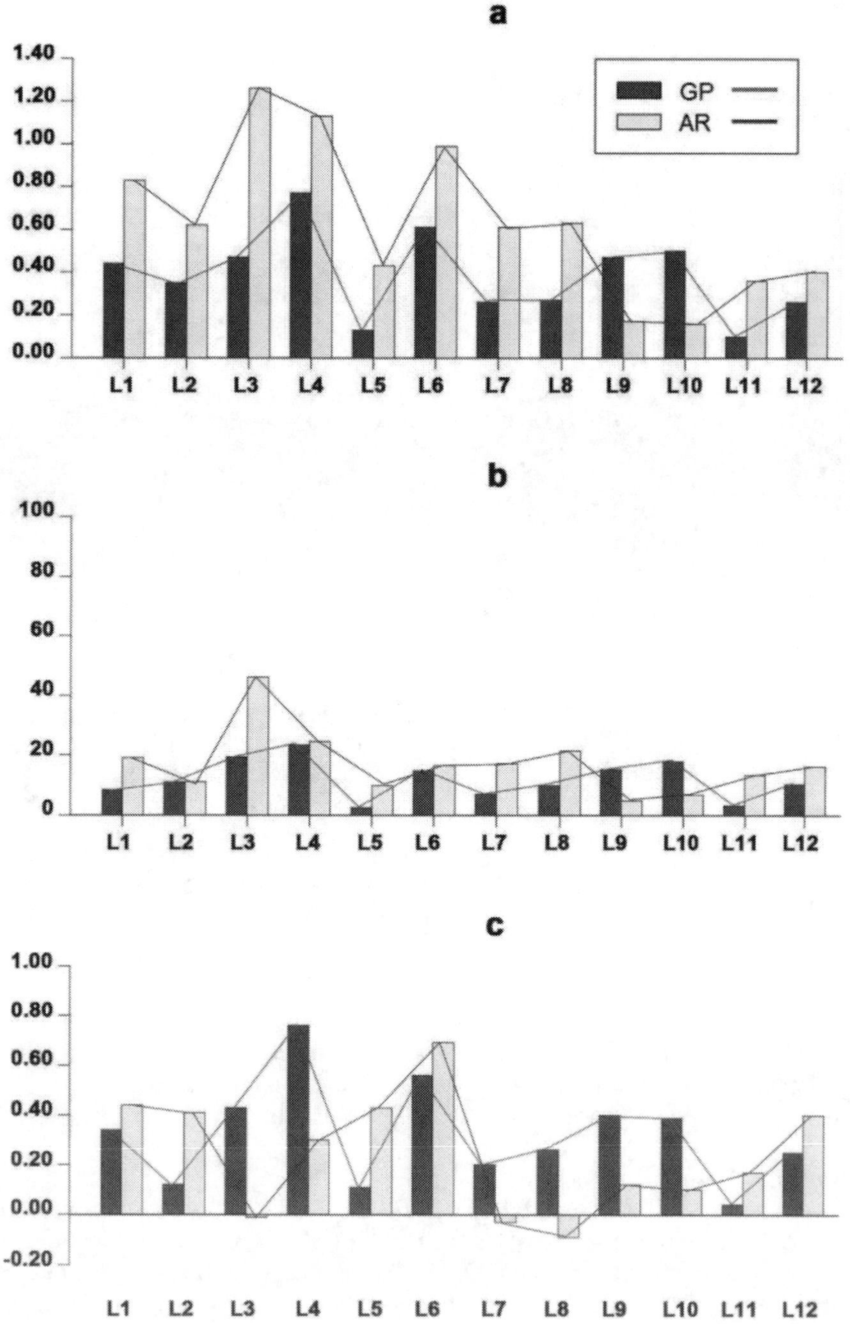

Figure 2. Statistical validation for (a) RMSE, (b) MAPE (in percentage), and (c) BIAS. Both units for RMSE and BIAS are in yy.

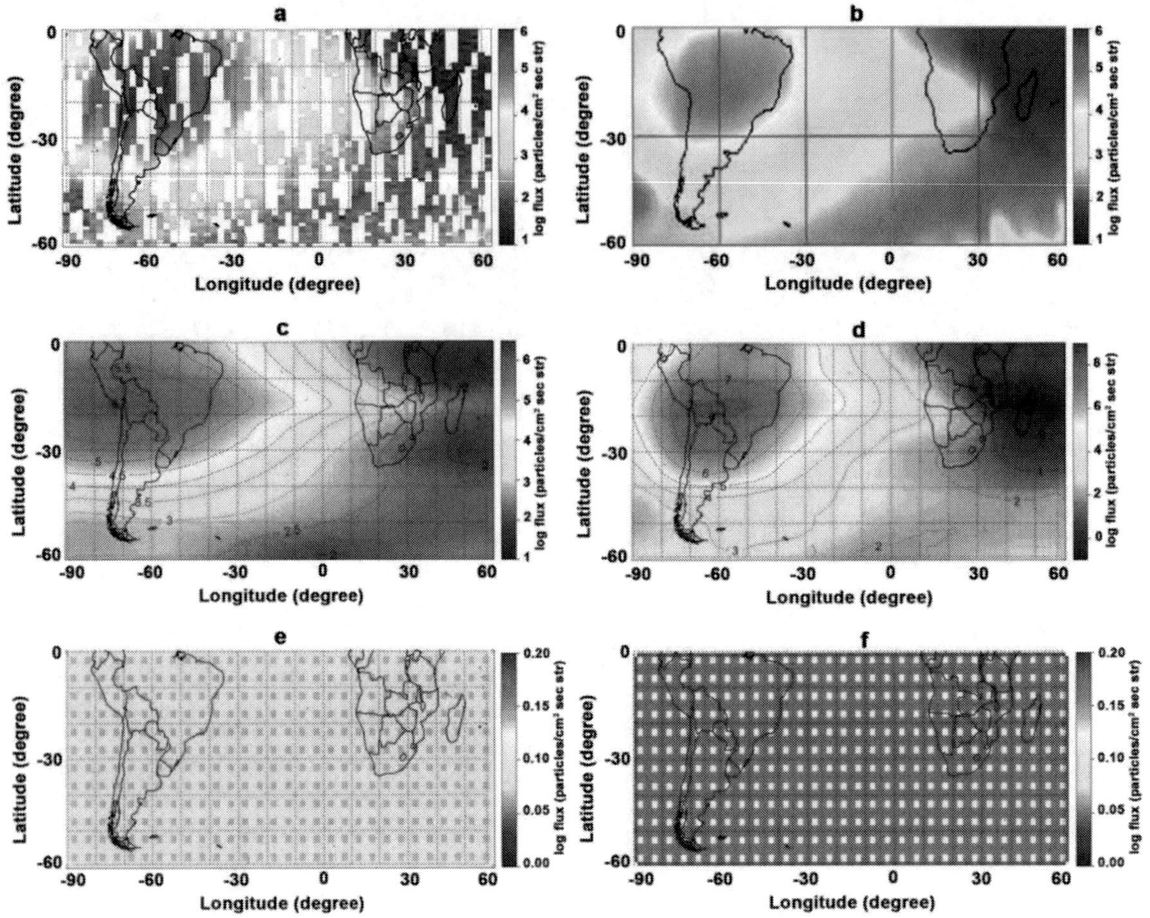

Figure 3. (a) Raw data plot of NOAA 15-18 on 31 August 2007 for mep0e1, (b) NOAA's flux distribution map of mep0e1 on 28-31 August 2007, (c) Kriging estimation on GP forecast result, (d) Kriging Estimation on AR forecast result, (e) Kriging variance on GP forecast result and (f) Kriging variance on AR forecast result.

4. Summary and Future work

This work has been successfully compared two HBST models to forecast trapped particle flux distribution on 31 August 2007 from 14-day input of NOAA's data. The GP obtained 88.02 % for forecasting accuracy weather the AR obtained 82.7 %. We assume that the less accurate of AR due to the AR temporal correlation parameter used in this work, that needs to be modified regarding to our data. Compared to our previous work, the employment of multiple satellites data in this work omits the missing values occurrence at validation points. In the future work, we propose to shrink the grid into 2°x2°size of longitude and latitude to reduce the average bias effect in the gridding process.

Acknowledgments

This work is supported by the Ministry of Science, Technology and Innovation (MOSTI) Malaysia under Science Fund 06-01-02-SF0808 grant. The authors also thank to NOAA for data used. Special thanks to Mr. Muhammad Marizal for his advice on R programming.

References

[1] Benton E R and Benton E V 2001 *Nucl. Instrum. Methods Phys. Res., Sect. B* **184** 255
[2] Gubby R and Evans J 2002 *J Atmos Sol-Terr Phy* **64** 1723
[3] Cressie N A and Wikle C K 2011 *Statistics for Spatio-Temporal Data* (New York: John Wiley & Sons)
[4] Suparta W, Gusrizal 2013 *paper presented at the 2014 International Conference on Science & Engineering in Mathematics, Chemistry and Physics (ScieTech 2014)* Jakarta 13-14 January 2014
[5] Banerjee S, Carlin B P and Gelfand A E 2004 *Hierarchical Modeling and Analysis for Spatial Data* (Boca Raton : Chapman & Hall/CRC)
[6] Sahu S K, Gelfand A E, and Holland D M 2007 *J. Am. Statist. Assoc.* **102** 1221
[7] Suparta W, Gusrizal, Mohd Alauddin M A and Ahmad N 2013 *Proceeding of the 2013 IEEE International Conference on Space Science and Communication (ICON SPACE)* Malacca 1-3 July 2013
[8] Sahu S K, Bakar K S, and Awang N 2013 *Technical Report* (Southampton: School of Mathematics University of Southampton)
[9] Bakar K S and Sahu S K 2013 *Technical Report* (Southampton: School of Mathematics University of Southampton)
[10] Furrer R, Nychka D, and Sain S 2013 *fields: Tools for Spatial Data* (http://CRAN.R-project.org/package=fields)

Twisted Skyrmion String

Miftachul Hadi[1,2,4,5], Malcolm Anderson[1], Andri Husein[3]
[1]Department of Mathematics, Universiti Brunei Darussalam, Negara Brunei Darussalam
[2]Physics Research Centre, Indonesian Insitute of Sciences (LIPI)
[3]Department of Physics, University of Sebelas Maret, Surakarta, Indonesia
[4]Department of Physics, Ulsan National Institute of Science and Technology, South Korea
[5]Institute of Modern Physics, Chinese Academy of Sciences, Lanzhou, China

E-mail: itpm.id@gmail.com

Abstract. We study nonlinear sigma model, especially Skyrme model without twist and Skyrme model with twist: twisted Skyrmion string. Twist term, mkz, is indicated in vortex solution. Necessary condition for stability of vortex solution has consequence that energy of vortex is minimum and scale-free (vortex solution is neutrally stable to changes in scale). We find numerically that the value of vortex minimum energy per unit length for twisted Skyrmion string is 20.37×10^{60} eV/m.

1. Introduction to Nonlinear Sigma Model

Nonlinear sigma model is a n-component scalar field theory where the field defines a mapping from space-time to a target manifold. A mapping here means a function from space-time to the target space [1].

By a nonlinear sigma model, we mean a field theory with the following properties [2]:

(1) The fields, $\phi(x)$, of the model are subjected to nonlinear constraints for all points $x \in \mathcal{M}_0$, where $\mathcal{M}_0$ is source (base) manifold, i.e. the spatial submanifold of the (2+1) or (3+1)-dimensional space-time manifold.

(2) The constraints and the Lagrangian density are invariant under the action of a global (space independent) symmetry group, G, on $\phi(x)$.

The Lagrangian density of a free (without potential) nonlinear sigma model on a Minkowski background space-time is defined as [3]

$$\mathcal{L} = \frac{1}{2\lambda^2}\, g^{ab}(\phi)\; \eta^{\mu\nu}\; \partial_\mu\phi_a\; \partial_\nu\phi_b \tag{1}$$

where $g^{ab}(\phi)$ is field metric, $\eta^{\mu\nu}$ is the Minkowski metric tensor, λ is a scaling constant with dimensions of $(\text{length}/\text{energy})^{1/2}$ and here ϕ is field. The nonlinearity is manifest in the field metric, $g_{ab}(\phi)$.

A special case of the nonlinear sigma model occurs when the target manifold is the unit sphere S^2 in R^3, i.e. $g^{ab}(\phi) = \delta^{ab}$. In case of $a = b$ then $\delta^{ab} = 1$, where δ^{ab} is Kronecker delta. The Lagrangian density (1) then becomes

$$\mathcal{L} = \frac{1}{2\lambda^2}\; \eta^{\mu\nu}\; \partial_\mu\phi \;.\; \partial_\nu\phi \tag{2}$$

where the dot (.) denotes the standard inner product on R^3, and the image of ϕ is S^2. Simple representation of ϕ (in case of time-dependent) is

$$\phi = \begin{pmatrix} \sin f(t,x)\ \sin g(t,x) \\ \sin f(t,x)\ \cos g(t,x) \\ \cos f(t,x) \end{pmatrix} \tag{3}$$

where f and g are scalar functions on the background space-time, with Minkowski coordinates $x^\mu = (t, x)$.

Substitute (3) into (2), then Lagrange density (2) becomes

$$\mathcal{L} = \frac{1}{2\lambda^2}(\eta^{\mu\nu}\ \partial_\mu f\ \partial_\nu f + [\sin^2 f]\ \eta^{\mu\nu}\ \partial_\mu g\ \partial_\nu g) \tag{4}$$

Associated Euler-Lagrange equations from $\mathcal{L}$ (4) are

$$\eta^{\mu\nu}\ \partial_\mu\partial_\nu f - (\sin f\ \cos f)\ \eta^{\mu\nu}\ \partial_\mu g\ \partial_\nu g = 0 \tag{5}$$

$$\eta^{\mu\nu}\ \partial_\mu\partial_\nu g + 2(\cot f)\ \eta^{\mu\nu}\ \partial_\mu f\ \partial_\nu g = 0 \tag{6}$$

2. $O(N)$ Nonlinear Sigma Model

The simplest example of nonlinear sigma models is the $O(N)$ nonlinear sigma model which consist of N-real scalar fields, ϕ^A, $A = 1, .., N$, having the Lagrangian density [2]

$$\mathcal{L} = \frac{1}{2}\ g^{\mu\nu}\ \frac{\partial\phi^A}{\partial x^\mu}\ \frac{\partial\phi^A}{\partial x^\nu} \tag{7}$$

where the scalar fields, ϕ^A, satisfy the constraint

$$\phi^A\phi^A = 1. \tag{8}$$

The Lagrangian density (7) is obviously invariant under the global (space independent) orthogonal transformations $O(N)$, i.e. the group of N-dimensional rotations [2]

$$\phi^A \to \phi'^A = O^A_B\ \phi^B. \tag{9}$$

One of the most interesting examples of $O(N)$ nonlinear sigma models due to its topological properties, is the $O(3)$ nonlinear sigma models in 1+1 dimensions, with the Lagrangian density [5]

$$\mathcal{L} = \frac{1}{2}\ \partial^\mu\phi \cdot \partial_\mu\phi \tag{10}$$

where $\phi = (\phi_1, \phi_2, \phi_3)$, due to $N = 3$, with the constraint $\phi \cdot \phi = 1$ and $\mu = 1, 2$.

3. Soliton Solution

Two solutions to these equations (5), (6), are

(i) A monopole solution, which has

$$\phi = \hat{\mathbf{r}} = \begin{pmatrix} x/\rho \\ y/\rho \\ z/\rho \end{pmatrix} \tag{11}$$

where $\rho = (x^2 + y^2 + z^2)^{1/2}$ is the spherical radius.

(ii) A vortex solution, which is found by imposing the "hedgehog" ansatz

$$\phi = \begin{pmatrix} \sin f(r)\ \sin(n\theta - \chi) \\ \sin f(r)\ \cos(n\theta - \chi) \\ \cos f(r) \end{pmatrix} \tag{12}$$

where $\theta = \arctan(x/y)$, n is a positive integer, and χ is a constant phase factor.

A vortex is a stable time-independent solution to a set of classical field equations that has finite energy in two spatial dimensions; it is a two dimensional soliton. In three spatial dimensions, a vortex becomes a string, a classical solution with finite energy per unit length [6]. Solutions of finite energy, satisfying the appropriate boundary conditions, are candidate soliton solutions [7].

Recall that there is in fact a family of vortex solutions

$$\sin f = \frac{2K^{1/2}r^n}{1 + Kr^{2n}} \tag{13}$$

or

$$\cos f = \frac{Kr^{2n} - 1}{Kr^{2n} + 1} \tag{14}$$

For each value of K where K is positive constant, there is a different vortex solution.

But, the mass per unit length

$$\mu = -\frac{4\pi n}{\lambda^2} \tag{15}$$

does not depend on K. (We use the same notation for energy per unit length and mass per unit length, due to equivalence of energy-mass as $E = mc^2$. Here, we take $c = 1$).

This means that the vortex solutions are what is called neutrally stable to changes in scale. As K change, the scale of the vortex changes, but the mass per unit length, μ, does not. Note that because of eq.(15), there is a preferred winding number, when n is a small as possible: $n = 1$. It means that for the vortex solution, the topological charge is just the winding number, n.

It can be shown that the topological charge is conserved, no matter what solution ϕ we have. So, topological charge is a constant, no matter what nonlinear sigma model we use. So long as

$$\phi = \begin{pmatrix} \sin f\ \sin g \\ \sin f\ \cos g \\ \cos f \end{pmatrix}. \tag{16}$$

Let us find f by solving the two equations of motion (5), (6). The function f satisfies the equation

$$r\frac{d^2 f}{dr^2} + \frac{df}{dr} - \frac{n^2}{r}\sin f\ \cos f = 0 \tag{17}$$

and that the solution satisfying the boundary conditions

$$f(0) = \pi \tag{18}$$

and

$$\lim_{r\to\infty} f(r) = 0 \tag{19}$$

is

$$\cos f = \frac{Kr^{2n} - 1}{Kr^{2n} + 1} \tag{20}$$

This is the vortex solution.

The energy density of a static (time-independent) field with Lagrangian density $\mathcal{L}$ (4) is

$$\begin{aligned} E &= -\mathcal{L} \\ &= -\frac{1}{2\lambda^2}\left[\eta^{\mu\nu}\,\partial_\mu f\,\partial_\nu f + (\sin^2 f)\,\eta^{\mu\nu}\,\partial_\mu g\,\partial_\nu g\right]. \end{aligned} \tag{21}$$

The energy density of the monopole solution is

$$E = \frac{1}{\lambda^2\rho^2} \tag{22}$$

and that the energy density of the vortex solutions is

$$E = \frac{4Kn^2}{\lambda^2}\frac{r^{2n-2}}{(Kr^{2n}+1)^2}. \tag{23}$$

Then the total energy

$$E = \int\int\int E\ dx\ dy\ dz, \tag{24}$$

of the monopole solution is infinite. But, that the energy per unit length of the vortex solutions

$$\mu = \int\int E\ dx\ dy = \frac{4\pi n}{\lambda^2} \tag{25}$$

is finite, and does not depend on the value of K.

This last fact means that the vortex solutions in the nonlinear sigma models have no preferred scale. A small value of K corresponds to a more extended vortex solution, and a larger value of K corresponds to a more compact vortex solution, as we can see by plotting f (or E) for different values of K and a fixed value of n (say, $n = 1$).

But, the value of the energy per unit length, μ, is the same for all these solutions, and so there is no natural size for the vortex solutions. It is for this reason that a Skyrme term is added to the Lagrangian density [4].

4. Skyrmion without Twist: Skyrme Model

We need to add Skyrme term to the Lagrangian to stabilize the vortex (which is neutrally stable to cylindrically symmetric perturbations). Original sigma model Lagrangian (in unit sphere) is

$$\mathcal{L}_1 = \frac{1}{2\lambda^2}\,\eta^{\mu\nu}\,\partial_\mu\phi\,.\,\partial_\nu\phi \tag{26}$$

Adding a Skyrme term to eq.(26), then eq.(26) becomes

$$\mathcal{L}_2 = \frac{1}{2\lambda^2}\,\eta^{\mu\nu}\,\partial_\mu\phi\,.\,\partial_\nu\phi - \underbrace{K_s\,\eta^{\kappa\lambda}\,\eta^{\mu\nu}(\partial_\kappa\phi\times\partial_\mu\phi)\,.\,(\partial_\lambda\phi\times\partial_\nu\phi)}_{\text{Skyrme term}} \tag{27}$$

Eq.(27) can be written in other expression by substituting (16) into (27). We get

$$\begin{aligned} \mathcal{L}_2 &= \frac{1}{2\lambda^2}\left(\eta^{\mu\nu}\,\partial_\mu f\,\partial_\nu f + \sin^2 f\ \eta^{\mu\nu}\,\partial_\mu g\,\partial_\nu g\right) \\ &\quad - K_s\left[2\sin^2 f\ (\eta^{\mu\nu}\,\partial_\mu f\,\partial_\nu f)\left(\eta^{\kappa\lambda}\,\partial_\kappa g\,\partial_\lambda g\right) - 2\sin^2 f\ (\eta^{\mu\nu}\,\partial_\mu f\,\partial_\nu g)^2\right] \end{aligned} \tag{28}$$

The Skyrme term becomes the second term on the right hand side of eq.(28). At this point, we need to write out Euler-Lagrange equations from $\mathcal{L}_2$ (28), i.e.

$$\partial_\alpha\left(\frac{\partial \mathcal{L}_2}{\partial(\partial_\alpha f)}\right) - \frac{\partial \mathcal{L}_2}{\partial f} = 0 \quad ; \quad \partial_\alpha\left(\frac{\partial \mathcal{L}_2}{\partial(\partial_\alpha g)}\right) - \frac{\partial \mathcal{L}_2}{\partial g} = 0 \tag{29}$$

Energy can be derived from $\mathcal{L}_2$ (28), as

$$E = \int\int \left\{ \frac{1}{2\lambda^2}\left[\left(\frac{df}{dr}\right)^2 + \frac{n^2}{r^2}\ \sin^2 f\right] - 2K_s\ \frac{n^2}{r^2}\ \sin^2 f\ \left(\frac{df}{dr}\right)^2 \right\} r\ dr\ d\theta \tag{30}$$

Let us define new variable

$$\overline{r} \equiv qr \tag{31}$$

where q is a constant. Then

$$\frac{df}{dr} = \frac{\partial f}{\partial \overline{r}}\ q \tag{32}$$

where $\partial\overline{r} = q\ \partial r$. So, the energy (30) can be rewritten using new variables as

$$E = \int\int \left\{ \frac{1}{2\lambda^2}\left[\left(\frac{\partial f}{\partial \overline{r}}\right)^2 + \frac{n^2}{r^2}\ \sin^2 f\right] - 2q^2\ K_s\ \frac{n^2}{\overline{r}^2}\ \sin^2 f\ \left(\frac{\partial f}{\partial \overline{r}}\right)^2 \right\} \overline{r}\ d\overline{r}\ d\theta \tag{33}$$

From (33), if we let $q \to \infty$ then the energy per unit length goes to $-\infty$. So, it is energetically favourable for the vortex to evolve , so that q increases. Then,

$$f(r) = f\left(\frac{\overline{r}}{q}\right) \tag{34}$$

and as $q \to \infty$ for fixed r, $\overline{r} \to \infty$ and the field evaporates to infinity. To fix this problem, we add a potential term, $K_v(1 - \underline{n}.\hat{\underline{\phi}})$, to Lagrangian density $\mathcal{L}_2$. So, we have

$$\mathcal{L}_3 = \mathcal{L}_2 + K_v(1 - \underline{n}.\hat{\underline{\phi}}) \tag{35}$$

where $\underline{n}$ is a direction of $\hat{\underline{\phi}}$ at $r = \infty$ (where, $f = 0$).

This $\mathcal{L}_3$ model is like Baby Skyrmion model [11] p.207, eq.(2.2). The kinetic term along with the Skyrme term are not sufficient to stabilize a baby Skyrmion, contrary to the usual Skyrme model. The kinetic term in $2+1$ dimensions enjoys (suffers from) conformal invariance and the baby Skyrmion can always reduce its energy by inflating (infinitely). Hence, one adds the mass term which limits the size of the baby Skyrmion. The usual Skyrme term of course prohibits the collapse of the soliton [12].

5. Skyrmion with Twist: Twisted Skyrmion String

Back to Skyrme model (28) and refer to eq.(16). Instead of choosing

$$g = n\theta - \chi \tag{36}$$

we choose

$$g = n\theta + mkz \tag{37}$$

where mkz is twist term. Then eq.(28) becomes

$$\mathcal{L}_2 = \frac{1}{2\lambda^2}\left[\left(\frac{df}{dr}\right)^2 + \sin^2 f\left(\frac{n^2}{r^2}+m^2k^2\right)\right] - 2K_s\ \sin^2 f\left(\frac{df}{dr}\right)^2\left(\frac{n^2}{r^2}+m^2k^2\right) \tag{38}$$

Twist is identified as direction of particle which rotates circularly around string (string can be imagined e.g. as a rod in z axis). The direction of twist can be clock-wise or counter clock-wise. There is a different value of "pressure" in clock-wise and counter clock-wise directions. Pressure is related with energy, it means that pressure is also related with mass, due to energy-mass relation [4].

Euler-Lagrange equation from $\mathcal{L}_2$ (28) with twist term (32), twisted Skyrmion string, is

$$\begin{aligned} 0 \ =\ & \frac{1}{\lambda^2}\left[\frac{d^2f}{dr^2}+\frac{1}{r}\frac{df}{dr}-\left(\frac{n^2}{r^2}+m^2k^2\right)\ \sin f\ \cos f\right] - 4\left(\frac{n^2}{r^2}+m^2k^2\right)K_s\ \sin^2 f\ \left(\frac{d^2f}{dr^2}-\frac{1}{r}\frac{df}{dr}\right) \\ & - 4\left(\frac{n^2}{r^2}+m^2k^2\right)K_s\ \sin f\ \cos f\ \left(\frac{df}{dr}\right)^2 \end{aligned} \tag{39}$$

6. Twisted Skyrmion String: Numerical Calculation and Result

Refer to eq.(39), let us rewrite Euler-Lagrange equation from $\mathcal{L}_2$, eq.(28), i.e. twisted Skyrmion string as

$$\begin{aligned} \frac{d^2f}{dr^2} \ =\ & -\frac{(\varepsilon+\zeta\ r^2)\ \sin f\ \cos f}{r^2+(\varepsilon+\zeta\ r^2)\ \sin^2 f}\left(\frac{df}{dr}\right)^2 - \frac{1}{r}\left(\frac{r^2-(\varepsilon+\zeta\ r^2)\ \sin^2 f}{r^2+(\varepsilon+\zeta\ r^2)\ \sin^2 f}\right)\frac{df}{dr} \\ & + \frac{n^2(1+\frac{\zeta}{\varepsilon}\ r^2)\ \sin f\ \cos f}{r^2+(\varepsilon+\zeta\ r^2)\ \sin^2 f} \end{aligned} \tag{40}$$

Eq.(40) can be solved numerically for different values of ε and ζ, starting with $f(0) = \pi$, $f'(0) = -a$ for different values of a. We need the numerical solution of (40) for calculating minimum energy of vortex.

Let us use Runge-Kutta fourth order method for solving (40). First, assume that

$$\frac{df}{dr} = v \tag{41}$$

$$\frac{d^2f}{dr^2} = \frac{dv}{dr} \tag{42}$$

Then

$$\begin{aligned} \frac{dv}{dr} \ =\ & g(r,\ f,\ v) \\ =\ & -\frac{(\varepsilon+\zeta\ r^2)\ \sin f\ \cos f}{r^2+(\varepsilon+\zeta r^2)\ \sin^2 f}\ v^2 - \frac{1}{r}\left(\frac{r^2-(\varepsilon+\zeta\ r^2)\ \sin^2 f}{r^2+(\varepsilon+\zeta\ r^2)\ \sin^2 f}\right)\ v \\ & + \frac{n^2(1+\frac{\zeta}{\varepsilon}\ r^2)\ \sin f\ \cos f}{r^2+(\varepsilon+\zeta\ r^2)\ \sin^2 f} \end{aligned} \tag{43}$$

Let us write down (41), (42) in iterative expression as

$$v_{i+1} = v_i + \frac{dr}{6}(k_1+2k_2+2k_3+k4) \tag{44}$$

Journal of Physics: Conference Series **539** (2014) 012008 doi:10.1088/1742-6596/539/1/012008

and

$$f_{i+1} = f_i + \frac{dr}{6}(l_1 + 2l_2 + 2l_3 + l_4) \tag{45}$$

with

$$l_1 = v, \qquad k_1 = g(r,\ f,\ v) \tag{46}$$

$$l_2 = v + \frac{dr}{2}k_1, \qquad k_2 = g(r + \frac{dr}{2},\ f + \frac{dr}{2}l_1,\ v + \frac{dr}{2}k_1) \tag{47}$$

$$l_3 = v + \frac{dr}{2}k_2, \qquad k_3 = g(r + \frac{dr}{2},\ f + \frac{dr}{2}l_2,\ v + \frac{dr}{2}k_2) \tag{48}$$

$$l_4 = v + dr\ k_3, \qquad k_4 = g(r + dr,\ f + dr\ l_3,\ v + dr\ k_3) \tag{49}$$

Eq.(44) is numerical solution of df/dr.

We obtain energy per unit length, μ, which can be derived from Lagrangian, $\mathcal{L}_2$, in eq.(28). Replace r with $\bar{r} = qr$ (q is scale factor, a number) then energy per unit length can be written as

$$\mu = 2\pi \int_0^\infty \left[\left(\frac{df}{d\bar{r}}\right)^2 + n^2\left(\frac{1}{\bar{r}^2} + \frac{\zeta}{\varepsilon\ q^2}\right)\sin^2 f + \left(\frac{df}{d\bar{r}}\right)^2\left(\zeta + \frac{\varepsilon\ q^2}{\bar{r}^2}\right)\ \sin^2 f \right] \bar{r}\ d\bar{r} \tag{50}$$

Eq.(50) can be written as

$$\mu = 2\pi \int_0^\infty \left[\left(\bar{r} + \zeta\ \bar{r}\ \sin^2 f + \frac{\varepsilon\ q^2}{\bar{r}}\ \sin^2 f\right)\left(\frac{df}{d\bar{r}}\right)^2 + n^2\left(\frac{1}{\bar{r}} + \frac{\zeta\ \bar{r}}{\varepsilon\ q^2}\right)\ \sin^2 f \right]\ d\bar{r} \tag{51}$$

From eq.(51) we are able to define

$$\eta \equiv \left(\bar{r} + \zeta\ \bar{r}\ \sin^2 f + \frac{\varepsilon\ q^2}{\bar{r}}\sin^2 f\right)\left(\frac{df}{d\bar{r}}\right)^2 + n^2\left(\frac{1}{\bar{r}} + \frac{\zeta\ \bar{r}}{\varepsilon\ q^2}\right)\sin^2 f \tag{52}$$

So, eq.(52) can be rewritten as

$$\mu = 2\pi \int_0^\infty \eta\ d\bar{r} \tag{53}$$

From chain rule, we have the relation that

$$\frac{\partial \mu}{\partial q} = \frac{\partial \mu}{\partial \bar{r}}\ \frac{\partial \bar{r}}{\partial q} \longrightarrow \frac{\partial \eta}{\partial q} = \frac{\partial \eta}{\partial \bar{r}}\ \frac{\partial \bar{r}}{\partial q} \tag{54}$$

Refer to (53) and (54), we can derive relation as below

$$\frac{\partial \mu}{\partial q} = 2\pi\ \frac{\partial}{\partial q}\int_0^\infty \eta\ d\bar{r} = 2\pi \int_0^\infty \frac{\partial \eta}{\partial q}\ d\bar{r} = 2\pi \int_0^\infty \frac{\partial \eta}{\partial \bar{r}}\ \frac{\partial \bar{r}}{\partial q}\ d\bar{r} = 2\pi \int_0^\infty \frac{\partial \eta}{\partial \bar{r}}\ r\ d\bar{r} = \frac{2\pi}{q}\int_0^\infty \frac{\partial \eta}{\partial \bar{r}}\ \bar{r}\ d\bar{r} \tag{55}$$

(Integration and differentiation operations are interchangeable, if their functions are continuous.) In order to evaluate the right hand side of eq.(55), we use identity relation

$$\frac{\partial}{\partial \bar{r}}(\eta\ \bar{r}) = \frac{\partial \eta}{\partial \bar{r}}\ \bar{r} + \eta \quad \longrightarrow \quad \frac{\partial \eta}{\partial \bar{r}}\ \bar{r} = \frac{\partial}{\partial \bar{r}}(\eta\ \bar{r}) - \eta \tag{56}$$

Then, we obtain

$$\frac{\partial \mu}{\partial q} = \frac{2\pi}{q} \int_0^\infty \frac{d\eta}{d\bar{r}}\ \bar{r}\ d\bar{r} = \frac{2\pi}{q} \int_0^\infty \left[\frac{\partial}{\partial \bar{r}}(\eta\ \bar{r}) - \eta\right]\ d\bar{r} = \frac{2\pi}{q}\ \eta\ \bar{r}|_0^\infty - \frac{2\pi}{q}\int_0^\infty \eta\ d\bar{r} \tag{57}$$

Substitute (52), (53) into eq.(57), we obtain

$$\begin{aligned} \frac{\partial \mu}{\partial q} &= \frac{2\pi}{q}\left[\left(\bar{r}^2 + \zeta\ \bar{r}^2\ \sin^2 f + \varepsilon\ q^2\ \sin^2 f\right)\left(\frac{df}{d\bar{r}}\right)^2 + n^2\left(1 + \frac{\zeta\ \bar{r}^2}{\varepsilon\ q^2}\right)\sin^2 f\right]_0^\infty \\ &\quad - \frac{1}{q}\ \mu \end{aligned} \tag{58}$$

Necessary condition for stability of vortex solution requires

$$\left.\frac{\partial \mu}{\partial q}\right|_{q=1} = 0 \tag{59}$$

From (58), (59), we obtain

$$0 = \frac{2\pi}{q}\left[\left(\bar{r}^2 + \zeta\ \bar{r}^2\ \sin^2 f + \varepsilon\ q^2\ \sin^2 f\right)\left(\frac{df}{d\bar{r}}\right)^2 + n^2\left(1 + \frac{\zeta\ \bar{r}^2}{\varepsilon\ q^2}\right)\sin^2 f\right]_0^\infty - \frac{1}{q}\mu \tag{60}$$

For $q = 1$, it has consequence that $\bar{r} \to r$ and $\mu \to \mu_{\text{min}}$. We obtain

$$\mu_{\text{min}} = 2\pi\left[\left(r^2 + \zeta\ r^2\ \sin^2 f + \varepsilon\ \sin^2 f\right)\left(\frac{df}{dr}\right)^2 + n^2\left(1 + \frac{\zeta\ r^2}{\varepsilon}\right)\sin^2 f\right]_0^\infty \tag{61}$$

where μ_{min} is minimum energy per unit length which fulfill the stability requirements. Relation between ζ parameter and minimum energy per unit length, μ, is shown **Figure 1** below:

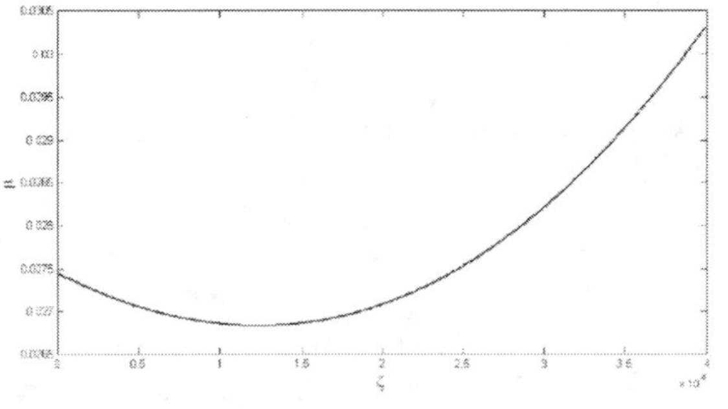

Figure 1. Result for numerical solution of (61) ζ versus μ, with parameters $a = 0.14$, $dr = 0.05$, $\varepsilon = 0.05$, $n = 1$, $r_\infty = 600$.

Here, we define J as a point of ζ where μ is minimum. We find that the value of J which is indicated by Matlab equal to 1.2400e-008. It means that the value of J is equal to 1.24×10^{-8}. We show both values, J and μ_{min}, in original form as shown by Mathlab in **Table 1** as below

Table 1. J and $\mu_{\min}$ relation

J	$\mu_{\min}$
`1.2400e-008`	0.0268

7. Discussion

Lagrangian density $\mathcal{L}_1$ (26) is the Lagrangian of the nonlinear sigma model. The vortex solutions are scale-free: the energy per unit length, μ, is independent of the width of the vortex. (We say that they the solutions are neutrally stable.)

Lagrangian density $\mathcal{L}_2$ (27), (28) are the Lagrangians with the Skyrme term added. The minimum energy per unit length occurs when the width of the untwisted vortex solution is infinite. So the vortex is unstable: it is energetically favourable for it to evaporate to infinity.

Lagrangian density $\mathcal{L}_3$ (35) is the Lagrangian with a stabilising potential added. The minimum energy per unit length occurs at a finite value of the width of the untwisted vortex solution. So, the untwisted vortex with this particular width is stable.

If instead of $\mathcal{L}_3$ we consider $\mathcal{L}_2$ and a twisted vortex solution (38), then for a weak field, the equation of motion of the twisted $\mathcal{L}_2$ vortex is the same as the equation of motion of the untwisted $\mathcal{L}_3$ vortex. We expect the twisted $\mathcal{L}_2$ vortex solutions to also be stable at a finite value of the width.

Necessary condition for stability of vortex solution has consequence that energy per unit length of vortex is minimum and scale-free. It means that vortex solutions are what is called neutrally stable to changes in scale. As scale factor change, the scale of the vortex changes, but the energy per unit length, does not.

Let us discuss about unit of energy in more detail. The mass per unit length, μ, of any cylindrically symmetric distribution of matter is usually quoted as a dimensionless quantity, meaning $G\ \mu/c^2$ where G is Newton's gravitational constant and c is the speed of light. (In relativity it is conventional to use "geometrized" units, in which $c = G = 1$, so the mass per length, μ, the energy per unit length, μc^2, and the dimensionless quantity, $G\mu/c^2$, are numerically the same.) So, μc^2 (now the physical energy per unit length) is normally quoted in units of c^4/G i.e. $c^4/G = 1.2 \times 10^{44}$ kg m/s^2 $= 1.2 \times 10^{44}$ J/m $= 7.6 \times 10^{62}$ eV/m.

We find graphically the value of ζ parameter and energy per unit length, μ, as shown in **Figure 1**. **Table 1** shows numerically that the value of minimum energy per unit length is 0.0268. It means that vortex minimum energy per unit length is $0.0268 \times 7.6 \times 10^{62}$ eV/m $=$ 20.37×10^{60} eV/m.

8. Acknowledgment

MH thank to Professor Malcolm Anderson for long patience and clear guidance. Thank also to Professor Eugen Simanek, Professor Edward Witten and Professor Wojtek Zakrzewski for fruitful discussions. Thank to Professor Yongmin Cho and Professor Pengming Zhang who withdraw my attention on topological objects in two-component Bose-Einstein condensates, which is static version of baby Skyrmion cosmic string. Mr Andri Husein for numerical works and fruitful discussions. Professor Muhaimin, Dr Irwandi Nurdin for kindly help. Department of Mathematics Universiti Brunei Darussalam, and Physics Research Centre LIPI for support and huge chances for doing this research. All kindly colleagues for their strong supports in various ways. Profound gratitude to beloved mother, Siti Ruchanah, and beloved Ika Nurlaila, for continuous praying and sincere love. Beloved Aliya Syauqina Hadi for purity and her naughtiness. This research is supported fully by Graduate Research Scholarships Universiti Brunei Darussalam (GRS UBD).

Journal of Physics: Conference Series **539** (2014) 012008 doi:10.1088/1742-6596/539/1/012008

9. References

[1] Zakrzewski W, private discussions
[2] Wospakrik H J, 2002 Harmonic Maps, SU(N) Skyrme Models and Yang-Mills Theories (Ph.D Thesis, University of Durham)
[3] Chen C.C. and Earnest T, April 7, 2010 Nonlinear Sigma Model, PHYS 583 (Urbana-Champaign: University of Illinois)
[4] Anderson M, private discussions
[5] Nonlinear Sigma Model (Wikipedia)
[6] Preskill J, 1987 Vortices and Monopoles (Elsevier Science Publishers B.V.)
[7] Manton N and Sutcliffe P, 2004 Topological Solitons (Cambridge: Cambridge Monographs on Mathematical Physics)
[8] Husein A, private discussions
[9] Bober W, T C Tay and Masory O, 2009 Numerial and Analytical Methods with MATLAB (CRC Press)
[10] Simanek E 28 Jan 2010 *arXiv: 1001.5061v1 [gr-qc]*
[11] Piette B.M.A.G, B.J. Schroers B.J. and Zakrzewski W.J. 1995 *Nuclear Physics* B **439** 205-235
[12] Gisiger T and Paranjape, M.B. 1996 *Physics Letters* B **384** 207-212
[13] Cho Y.M, Khim H and Zhang P.M. 2005 *Physical Review* A **72** 063603

The Importance of Site Selection for Radio Astronomy

Roslan Umar[1,2], Zamri Zainal Abidin[1], Zainol Abidin Ibrahim[1]

[1] Department of Physics, University of Malaya, 50603, Kuala Lumpur, Malaysia
[2] Astronomy Unit, Faculty of Islamic Contemporary Studies, Universiti Sultan Zainal Abidin, 21300, Kuala Terengganu, Malaysia

E-mail: roslan@unisza.edu.my

Abstract. Radio sources are very weak since this object travel very far from outer space. Radio astronomy studies are limited due to radio frequency interference (RFI) that is made by man. If the harassment is not stopped, it will provide critical problems in their radio astronomy scientists research. The purpose of this study is to provide RFI map Peninsular Malaysia with a minimum mapping techniques RFI interference. RFI mapping technique using GIS is proposed as a tool in mapping techniques. Decision-making process for the selection requires gathering information from a variety of parameters. These factors affecting the selection process are also taken account. In this study, various factors or parameters involved such as availability of telecommunications transmission (including radio and television), rainfall, water line and human activity. This study will benefit radio astronomy research especially in the RFI profile in Malaysia. Keywords: Radio Astronomy, Radio Frequency Interference (RFI), RFI mapping technique : GIS.

1. Introduction

Optimum coverage for radio astronomy observatory is important part need to consider. However interference from human activity are the main threats to radio astronomers. This due to the rapid urbanization result of the growth in industrial and commercial sector in the world has increased the demand of usage radio frequency. Radio Frequency Interference mapping using GIS is useful technique in order to identify interference. This technique helps to find and monitor spectrum occupancy in identifying and recognizing the most intense sources of interference, particularly from terrestrial transmitters. This paper demonstrates the potential of GIS and MCDA in selecting a suitable sites for a radio astronomy observatory in Peninsular of Malaysia. There are several factors in geographical and anthropogenic such as population density, contour shielding, road network and rainfall. This mapping technique is a tool in recognizing and identifying the most intense sources of interference. Parameters required were generated using data in GIS mode. This study offers tough, precise, cost a time effective procedure for preliminary selection for a radio astronomy observatory. However for the practical and final stage, site testing for RFI measurement need to do in order to see the profile of interference.

Journal of Physics: Conference Series **539** (2014) 012009 doi:10.1088/1742-6596/539/1/012009

2. Related Research

Site selection studies are very important for decision makers to proposed where to build an astronomical observatory especially in radio astronomy with a greatest efficiency. However, site selection not as easy as we think, it is a complex problem and involved multi criteria. In this selection there are two step must be taken namely site selection and the site testing. however in this [a[er we only do the site selection using several parameter interested. Drobne and Lisec, 2009 proposed a decision analyses involved systematic procedures in order to solve complicated problems namely MCDA. MCDA integrated with GIS a powerful tool with its ability to settle complex and various amount of data obtained from different sources offering many benefit such as cost effective, efficient and rapid management especially for the site selection problem. GIS is the collection, processing, storage [1]. There are many studies on the site selection for different purposes have been performed by using MCDA integrated with GIS such in order to improve the efficiency of dangerous and harmful species monitoring [2], looking for best site in optical astronomical observatory [3], flood risk assessment which is a case study of Huaihe River Basin in China [4], used in siting sanitary landfill in Northern Cyprus [5], to identify locations for ecotourism in Qeshm Island, Iran [6], using in the earthquake proof division of construction site [7], fire sites planning in China [8], used in environmental quality monitoring.

The major elements involved in the spatial multi-criteria analysis are shown in Figure 1 [9]. Figure 1 presents a three-stage hierarchy of intelligence, design, and choice to represent the decision making process. In the intelligence phase, data are acquired, processed, and exploratory data analysis is performed design phase usually the involves formal modeling/GIS interaction in order to develop a solution set of spatial decision alternatives. Integration of decision analytic techniques and GIS functions is critical for supporting the design phase. The choice phase involves selecting a particular alternative from those available. In this phase, specific decision rules are used to evaluate and rank alternatives. The three stages of decision making do not necessarily follow a linear path from intelligence, to design, and of choice [9].

3. The Study Area : Peninsular of Malaysia

Malaysia has been selected as the study area in order to select the best site which is pinpointed as good a possible site to build radio telescope in Peninsular of Malaysia. Malaysia can be divided into three main regions: Peninsular Malaysia, Sabah and Sarawak. Sabah and Sarawak also called Borneo. Approximate location of these three regions are given below. Peninsular : $06°45'$ and $01°02'$ N latitude and $99°40'$ and $104°20'$E longitudes, Sabah : $04°00'$ and $07°00'$ N latitude and $115°20'$ and $119°20'$E longitudes and Sarawak : $00°50'$ and $05°00'$ N latitude and $109°35'$ and $115°40'$E longitudes.

The Fig. 1 neighboring countries of Malaysia are Thailand and Brunei on the north and Singapore and Indonesia on the south (Fig. 2). The capital of Malaysia is Kuala Lumpur. The population stands at over about 35 million. Peninsular Malaysia borders by Thailand, Indonesia and Singapore however Borneo border with Brunei and Indonesia. In this study we only concentrate to the Peninsular of Malaysia. This country is located near the equator and experiences a tropical climate. This country also free from earthquake and volcano as well, this give an advantage of location of observatory. The country is generally warm throughout the year with temperatures ranging from $21°$ to $32°C$ in the lowlands. This can however be as low as $16°C$ in the highlands. Annual rainfall is from about 164 cm - 241 cm but sometime heavy at 250 cm. On the rainy day thunder and lightning often accompany the heavy downpour which normally lasts for about an hour or two. The high humidity level at 80 percent throughout the year favors light and sweat-absorbent material like cotton for drying season. In general, Malaysia has two distinct seasons. The dry season occurs during the southwest monsoon from May until September. The Northeast monsoon brings the rainy season to the country during

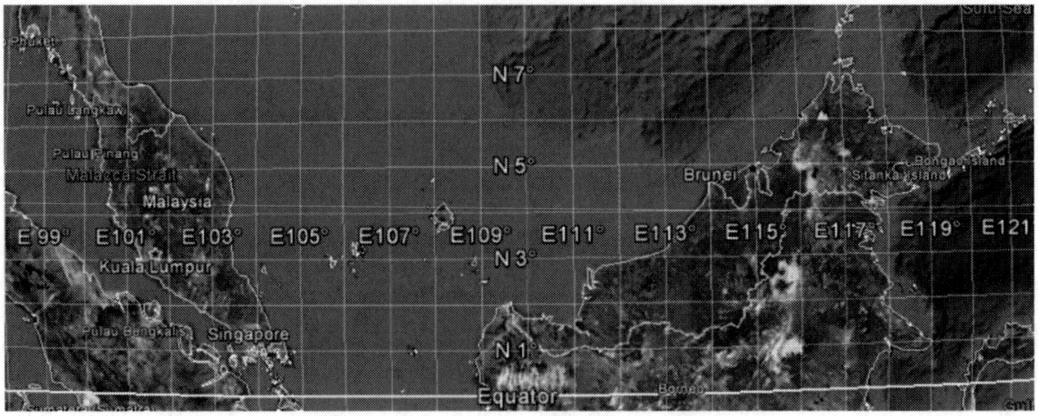

Figure 1. Location of study area

mid-November until March. The Peninsular Malaysia is divided into the east coast by the Main Range, know as Banjaran Titiwangsa which runs from the Malaysia- Thailand border in the north to the southern state of Negeri Sembilan. Geographically, East Malaysia is rugged, with a series of mountain range encompassing the interior regions of both states.

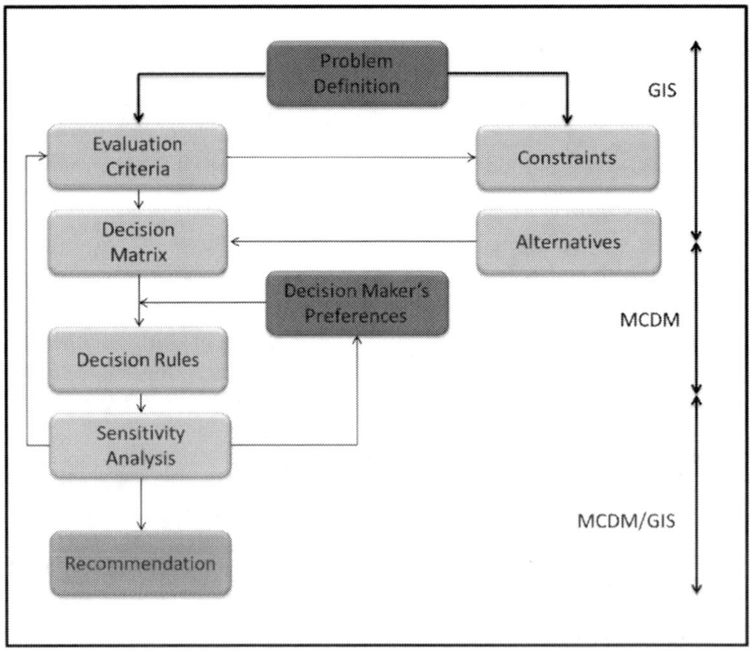

Figure 2. The flow chart of the process steps followed in GIS

4. Method

Multi-criteria decision analysis (MCDA) is a popular and useful technique that explicitly considers multiple criteria in decision-making environments. If one need to make a choice which deal with multiple conflicting criteria, he has to consider the factors that may influence or affect the final result. He has also need to evaluate the ability of various activities and then the corresponding attributes or indicators are identified to fulfil a given objective. The indicators are often defined based on scores achieved, rank and weightage. This method is capable to support decision makers facing such problems.

In this study, the MCDA technique is the best, well known structured approach to be practiced. As a result, the region in all around peninsular Malaysia will be ranked out from the lowest RFI(the best site) to the highest RFI(the most inappropriate place to consider for observation).

The step-by-step in methodology process of RFI map is shown in Fig. 3. Once the parameters have been set up, the Multi-Criteria Decision Analysis(MCDA) plays the role to produce the predicted RFI map. In this analysis, three main step are involved: determine the threshold value for each of the parameters, combine the data to produce a map for each parameter, and perform an Analytical Hierarchy Process(AHP) method. The MCDA integrated with the GIS process is illustrated in.

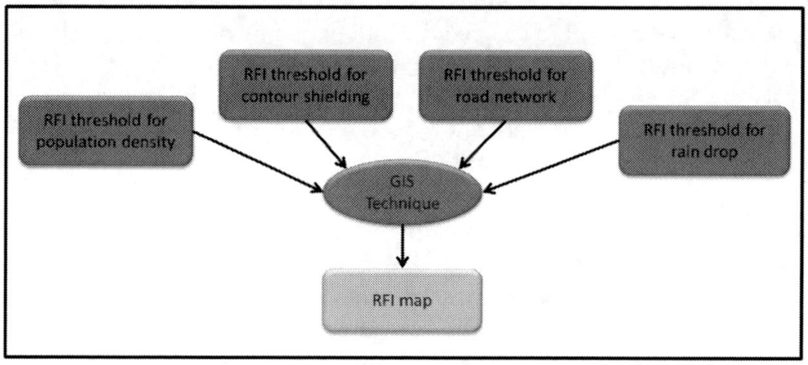

Figure 3. The flow chart of the process steps followed in methodology

4.1. Source of datas

The data was requested directly from various agencies such as Malaysia Geo-Spatial Data Infrastructure (MyGDI), Department of Statistics Malaysia, Malaysian Meteorological Department, the Malaysian Communications and Multimedia Commission (MCMC) and the Department of Malaysia Map and Survey data including spatial and attribute data. Spatial data is a map of Peninsular Malaysia, which we get from the Department of Mapping and Measurement Malaysia. Other sources of support also in the spatial data, including topology maps to high places in Malaysia and GIS maps obtained from previous studies of the Faculty of Social Sciences and Humanities (FSSS), University of Malaya.

4.2. Processing of data sets using GIS

In this section we will split data with two sets of anthropogenic and geography. In anthropogenic data, we have the population density and road. However, we have a set of geographic data and raindrop contour. A set of data associated with human activity for example distance to roads and population density are the anthropogenic. They directly affect the quality of the observations, this is the same as in the case of optical observation. Another important parameter is the geographical situation near the observatory. As we know that the expensive equipment in the observatory should not be built in areas that are not safe in terms of geological conditions such as landslides and earth quake zone. Geography can help to reduce the level of interference as we know that the contour and raindrops become one of the options parameter by ALMA and SKA project to protect from interferens for example.

Journal of Physics: Conference Series **539** (2014) 012009 doi:10.1088/1742-6596/539/1/012009

4.3. Anthropogenic data sets

4.3.1. Population density data There is a parameter which is very related to human activity is population density. This population density represent the residential and commercial. This population density data was digitized as raster data and reclassify from 0 to 100. value of 100 represent best site meanwhile 0 least suitable site.

4.3.2. Road network data Another important variable need to take care of was road network. Road information was present the usage of vehicle and develop area as well. When we discussing the road network, it also present the type of vehicle usage on the highway used. value of 100 represent best site meanwhile 0 least suitable site.

4.4. Geographical data sets

4.4.1. Contour shielding data To locating of radio telescopes to avoid spectrum pollution most of radio observatories are often placed in valleys to further shield them from ground interference. So that altitude is a one factor in the site selection of radio observatory. Most of radio observatory in the world especially for high frequency study located at highest site. Meanwhile radio observation for low frequency normally located at lower in altitude. value of 100 represent best site meanwhile 0 least suitable site.

4.4.2. Rainfall data Rainfall is one of important variable to radio propagation. It will affect penetrate of wave radio propagation where the water occupied in atmosphere will reflect or refract and attenuate as well the signal penetrate in the atmosphere. We need to find driest place in peninsular of Malaysia. In this study, rain fall range in Malaysia is from $164cm^3 - 241cm^3$. Value of 100 represent best site meanwhile 0 least suitable site.

4.5. The analytic hierarchy process

After representing the criteria map layers in raster data format and rescaling the values between 0 and 100, all these layers were combined into a GIS. AHP was used for obtaining the weights connected with criteria map layers. In generally, AHP is based on pairwise comparisons for each data layer using numerical grade in an absolute scale of numbers (Saaty, 1980 and Saaty, 2008) and provides a very simple hierarchy to solving complicated problem. In AHP, single number value within the range of 19 was assigned as a degree of importance between two criteria.

These numbers were assigned to each pairwise by decision makers who are experts in astronomical observations, geosciences, meteorology and urban planning. These experts collaborated with each other to complement their knowledge and skills for deciding on suitable astronomical observatory site. In order to construct an AHP matrix which contains grade of importance between pairwise data, pairwise comparison is performed. The intensity of importance scale has a range between 1 and 9. If the compared data have equal importance, then the value in the matrix is one. When the reference data has extremely more important than the compared data, the value of the comparison is given as the highest number in AHP matrix which is nine.

5. Result

Each comparison grade is written on the pairwise comparison matrix (AHP matrix). In pairwise comparison matrix, matrix element is formed as $a_{ij} = (1/a_{ji})$. Due to the fact that column and

row data layers are equal to each. other; diagonal values on this matrix are equal to one ($a_{ii} = 1$, $a_{jj} = 1$, $a_{nn} = 1$ etc.). The grading values of all pairs are normalized to one using maximum absolute eigenvalue (Eqs. 1 and 2).

$$\lambda_{\max}\frac{1}{n}\sum_{w_i}^{n}\frac{(AW)_i}{w_i} \tag{1}$$

referring to Eq.1 base on report by [?]

$$AW = \begin{bmatrix} a_{l_1} & \cdots & a_{l_n} \\ \vdots & \ddots & \vdots \\ a_{n_1} & \cdots & a_{nn} \end{bmatrix} \cdot \begin{bmatrix} w_1 \\ \\ w_i \end{bmatrix} \tag{2}$$

Where A is pairwise comparison matrix. W and wi (i =1,2, ...,n) stand for corresponding eigenvector of kmax and the weight value for ranking, respectively. At the beginning of the processes, pairwise comparison matrix was constructed using 4 criteria, which are previously created from different data sources (Table 2). After generating matrix, weights were computed for each parameter. kmax value was obtained as 4.055 in this study. Before using these weights, consistency of the pairwise comparison matrix should be confirmed that is consistent or not by using consistency index (CI, Eq. 3)

$$CI = \frac{\lambda_{\max} - 1}{n - 1} \tag{3}$$

Where $\lambda_{\max}$ and n are the consistency index, the largest (principle) eigenvalue of the matrix and the dimension of the matrix (number of input parameters), receptively. CI value was calculated as 0.0184 in this study. In order to test the consistent of matrix we construct, the consistent ratio (CR) value was calculate base on value studied by Saaty 1980 (cite Saaty). CR formula given as equation below (Eq. 4);

$$CR = \frac{CI}{RI} \tag{4}$$

Where CI is the consistent index and RI is random Inconsistency (refer paper Saaty 1980 and Tran 2007). After we have obtain the value is less than 0.10 for example, the value of consistency generated by pairwise comparison is reasonable (Saaty 2008). We noted that RI value for n = 4 was 0.90, this value given by Saaty and Tran 2007. So that from the formula we CR obtained 0.0204 in this study. Therefore our aur pairwise for matrix is acceptable. if the value computed larger that 0.1 means that pairwise comparison should be renewed to get the value under limit (0.1). Table 1 show all value computed weight for all variables, CI, $\lambda_{\max}$, RI and CR value computed in this paper. In the final stage, each data layer that we analysis will combined by using the calculated weight coefficients to obtain the best sites result for radio observatory using GIS tools.

Table 1. AHP pairwise comparison matrix and computed weights. * A: Population density; B: Contour shielding, C: Road network and D: Rainfall

	A	B	C	D	Weight
A	0.052	0.021	0.045	0.081	0.0503
B	0.263	0.195	0.045	0.285	0.197
C	0.105	0.195	0.090	0.0633	0.113
D	0.210	0.586	0.545	0.5701	0.478

The resultant map that indicates candidate sites was produced after performing the dense computation for obtaining weights of corresponding criteria and multiplying each criterion with

Journal of Physics: Conference Series **539** (2014) 012009 doi:10.1088/1742-6596/539/1/012009

these weights using AHP and GIS. The best possible sites that were obtained by applying a threshold value to select the areas having highest scores are presented in Fig. 4. In this figure, the suitable areas in terms of meteorological, geographical and anthropogenic conditions, which were extensively located in western and eastern part of the study area, are illustrated in red1 colors (Fig. 4). The best site could be low in RFI is map represent by grey and yellow. Moderate RFI could be present by green colour and High in RFI represent by blue and red colour as ilustrated in Fig. 4.

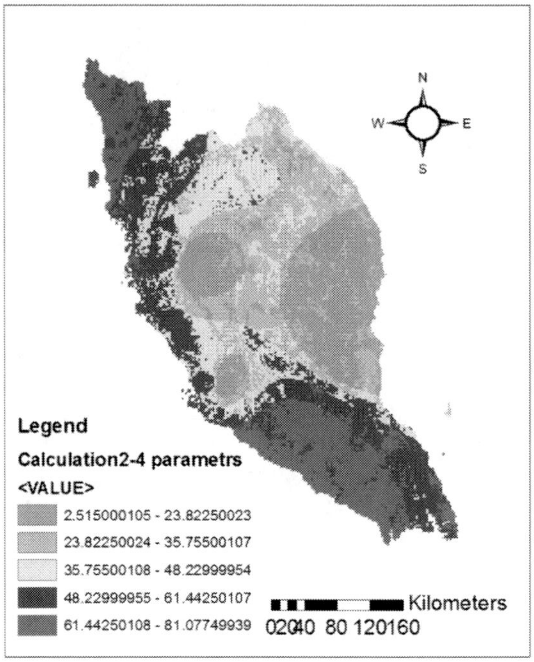

Figure 4. After consider four parameters

6. Conclusions

One of the most important factors that affect the resultant map is the construction of pairwise comparison matrix for AHP. The values on that matrix should be given by the expert who has an idea about observatories and their locations, geographical and meteorological conditions. The expert must know the effect and importance of each criterion for astronomical observatory site selection. If the values inside the matrix are not reliable, then the weights of the parameters cannot be correctly computed and the final map may not represent the accurate places. In this study, the meteorological data sets, which include cloud cover and precipitate water factors, are important as mentioned before. The remotely sensed data for meteorological data sets (i.e. MODIS data) provide spatial.

This study provides an example application of the MCDA, remote sensing and GIS technologies for the purposes of site selection for astronomical observatory. By handling and evaluating complex criteria, our approach proves to be an efficient and powerful method for astronomical observatory site selection studies. This study will be followed by astronomical site testing and meteorological observations on the spot at those selected sites for a final decision.

The developed approach in this study for the selection was implemented in Antalya province of Turkey. However, it can be applied efficiently to any city or region, in which an astronomical observatory site construction is planning.

Besides of the site selection, the other benefit of this study can be helpful to find best season or month within a year for astronomical observations If more time series of meteorological data is available, then suitable astronomical observation periods can be computed statistically for future work.

Stronger than typical radio astronomical source can be refer in [10]. Another example of the problems that can arise is provided by the iridium mobile communications system, which has a group of satellites transmitting signals to every point on the surface of the Earth. Unfortunately in this case, there is some leakage into the passive band about $1612MHz$, with signal levels up to 10^{11} times as strong as signals than sources from the early universe [11].

For the future research not only population density (human generated RFI) but another factor or parameter such as telecomunication transmitters and airplane route could be important in order to find suitable place in least RFI [12]. Radio windows focus also should be extend to the high frequency such as S-Band and K-band not only in hydrogen line located at L-band [13].

6.1. Acknowledgments

Authors wishing to acknowledge assistance or encouragement from colleagues, special work by technical staff or financial support from organizations should do so in an unnumbered Acknowledgments section immediately following the last numbered section of the paper. The command `\ack` sets the acknowledgments heading as an unnumbered section.

References

[1] Qian, Ru 2012 *Proceedings of the 2012 International Conference of Modern Computer Science and Applications*
[2] Chen H, Weng Q, Chi M, Qiu R, and Zhao J 2013 *Construction of Monitoring System of Dangerous and Harmful Species of Import Taiwan Fruits and Vegetables Based on GIS* (Computer and Computing Technologies in Agriculture VI: Springer)
[3] Koc-San D, San B, Bakis V, Helvaci M & Eker Z 2000 *Advances in Space Research.* **52** 1
[4] Liu, JF and Li, Jing and Liu, Jing and Cao, Ru-yin 2008 *J Nat Disasters.* **17**
[5] Kara Can and Doratli Naciye 2012 *Waste Management & Research.* **30** 9
[6] Dashti S, Monavari S M, Hosseini S M, Riazi B & Momeni M 2013 *Life Science Journal.* **13**
[7] Lou Ke-Cheng and Ding, Bo-Yang and Li, Ang Shi-Jun 2010 *West-China Exploration Engineering.* **22** 9
[8] Ai, WEI and Han-lun, LI and Qi, LIU and Jing-yi, CHEN and Yi-jiao, CUI 2011 *Procedia Engineering.* **11**
[9] Malczewski J 1999 *GIS and multicriteria decision analysis* (John Wiley & Sons)
[10] Abidin Z Z, Ibrahim Z A, Bahari S.; Syed R & Anuar N K 2009 *New Astronomy.* **14**
[11] Ekers R & Bell J 2000 *The Future of Radio Astronomy: Options for Dealing with Human Generated Interference Preserving the Astronomical Sky IAU Symposium 196*
Umar R, Abidin Z Z, Ibrahim Z A, Rosli Z & Noorazlan N 2014 *Research in Astronomy and Astrophysics (RAA).* **14**
[13] Abidin Z, Umar R, Ibrahim Z, Rosli Z, Asanok K & Gasiprong N 2013 *Publications of the Astronomical Society of Australia, Cambridge Univ Press,.* **30**

Journal of Physics: Conference Series **539** (2014) 012010 doi:10.1088/1742-6596/539/1/012010

Radiation Dose Analysis of Galactic Cosmic Ray in Low Earth Orbit/Near Equatorial Orbit

W Suparta[1] and S K Zulkeple[1]
Space Science Centre (ANGKASA), Institute of Climate Change, Universiti Kebangsaan Malaysia, 43600, Bangi, Selangor Darul Ehsan, Malaysia

E-mail: wayan@ukm.my

Abstract. Space environment contained harmful radiation that posed risk to spacecraft orbiting the Earth. In this paper, we looked into radiation doses caused by galactic cosmic ray (GCR) towards satellites orbiting in low earth orbit (LEO) near Earth's equator (NEqO) and compared them with doses caused by solar energetic particles (SEP) and trapped particles to determine the damage level of GCR. The radiation doses included linear energy transfer (LET) and non-ionizing energy loss (NIEL) through a 1mm gallium arsenide (GaAs) planar geometry by using Space Environment Information System (SPENVIS) method. The orbital data followed Malaysian Razaksat satellite at 685km altitude and 9° inclination during selected solar minimum and solar maximum from solar cycles 21 to 24. We found that trapped particles gave the highest LET and no SEP was detected in SPENVIS. The LET values tend to be higher during solar minimum for trapped particles and GCR, corresponding to their anti-correlated fluxes with the solar activity. However, the NIEL values for GCR in solar cycle 23 did not follow the anti-correlation pattern.

1. Introduction

Many spacecraft located in low earth orbit (LEO) have lower radiation risks but there is still no guarantee that they are safe from space hazards when space conditions can be unpredictable. At altitudes starting from LEO and beyond, the highly energized space particles could affect microelectronics and photonics as a result from secondary productions [1].

Based on our previous study on spatial analysis of galactic cosmic ray (GCR) [2], we discovered that orbit located at low earth near the equator (LEO/NEqO) is dominated by high energy GCR of more than 1 GeV while trapped particles dominated the lower energy range with protons less than 400 MeV and electrons less than 1 MeV, at the same time, no solar energetic particles (SEP) was found. With the exception of SEP, it is without a doubt that GCR and trapped particles played a significant role in LEO/NEqO orbit. Therefore, in this study, we decided to look into the effects of these highly charged particles by considering ionizing and non-ionizing radiation doses caused. The doses included linear energy transfer (LET) and non-ionizing energy loss (NIEL) towards gallium arsenide (GaAs) material which is one of the materials that made up the Malaysian Razaksat satellite to determine its displacement damage. This is an important aspect because atomic displacement reduced efficiency in solar cells, degrading resolution in solid-state detectors and demeaning charge transfer in charge coupled devices, all of which are crucial in a satellite [3].

Journal of Physics: Conference Series **539** (2014) 012010 doi:10.1088/1742-6596/539/1/012010

LET is the energy loss by an ionizing particle over its path length within a sensitive volume and is a quantity to best define heavier nuclei and mono-energetic ions such as proton [4]. Considering that single event upset (SEU) is an effect from direct ionization of charged particles, thus LET becomes a suitable parameter for this study.

NIEL, on the other hand, is the energy a particle conveys to a solid via mechanisms other than ionization and is the cause of atomic displacement. It accounts for Coulomb interactions as well as nuclear elastic and non-elastic reactions. Furthermore, this metric is important when SEU is produced through nuclear reactions in which by then LET metric becomes invalid [5].

2. Methodology

The tool used was Space Environment Information System (SPENVIS) as it incorporated many models to describe the space environment and its effects towards various materials [6]. To use this tool, users have to first generate orbit. Malaysian Razaksat satellite orbital data is used to represent LEO/NEqO orbit with its 685km altitude and 9° inclination. The overview of the methodology is simplified in Figure 1.

We used ISO 15390 model to obtain GCR energies and fluxes because it considers the particles with respect to variations in the solar activity and the sun's magnetic field throughout 22-year cycles by accounting the latest solar minimum data which is May 1996 [7]. JPL model is chosen to collect SEP data based on its data set that was taken only during 7 active years of the solar cycles by assuming that no significant proton fluences occur during the quiet period [8]. Meanwhile, models selected for trapped particles were AP8 for proton and AE8 for electron due to the fact that they cover a wide range of energies for the whole region of the Earth's radiation belt [9].

Once the orbit was generated, the data on the energetic charged particles were selected to an average of 30 days during solar minimum and solar maximum for solar cycle 21 until part of solar cycle 24 under a quiet magnetosphere. The period chosen for solar minimum and maximum for each of the solar cycle is tabulated in Table 1. The fluences obtained were then simulated into Geant4 Radiation Analysis for Space (GRAS) tool for the calculations of LET and NIEL. The geometry in this tool was set as 1mm GaAs planar.

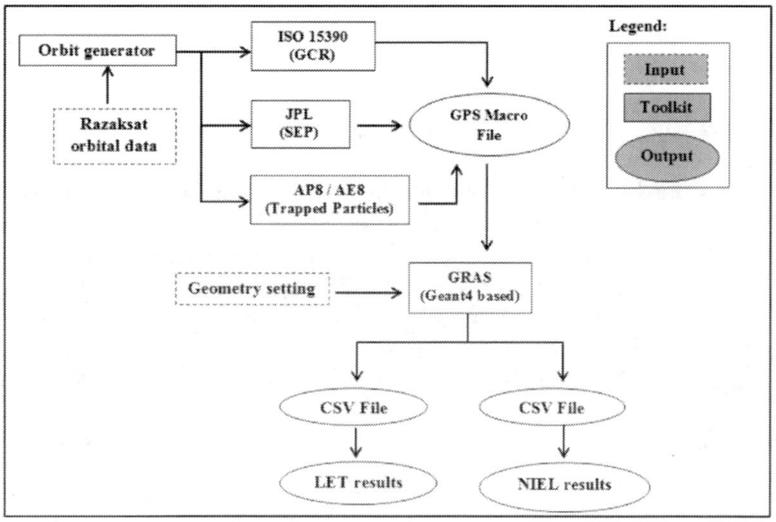

Figure 1. Block diagram of procedures to generate radiation dose calculations in SPENVIS using 30 days average data during selected solar minimum and solar maximum from solar cycle 21 to solar cycle 24

Journal of Physics: Conference Series **539** (2014) 012010 doi:10.1088/1742-6596/539/1/012010

Table 1. Period data taken to represent solar minimum and solar maximum for four recent solar cycles

Solar Cycle	Solar Minimum	Solar Maximum
21	August 1986	December 1979
22	May 1996	June 1989
23	March 2007	April 2000
24	January 2008	December 2013

3. Results and Discussion

Comparison between GCR with SEP and trapped particles are presented in terms of LET and NIEL values through a simple 1mm GaAs planar geometry. The results for LET and NIEL for four recent solar cycles are shown in Table 2 and 3 respectively.

As shown in Figure 2, LET for trapped protons were considerably higher than GCR for both solar minimum and solar maximum in all the four solar cycles. This was no surprise as trapped particles populated the Earth's radiation belt below 20,000 km altitude so the resulting LET for trapped protons in this study were twice higher than trapped electrons. Moreover, in the Earth's radiation belt, there is a phenomenon where it would be the closest to the surface, called South Atlantic Anomaly (SAA) in a region off the coast of Brazil. The flux of trapped protons is the highest here and this could contribute to their greater LET value.

The difference between solar activities can be seen in the figure whereby it made a significant influence towards the LET value for trapped protons. Their LET is greater during solar minimum when solar wind had the most intense speed. The population of trapped protons is believed to be resulted from Cosmic Ray Albedo Neutron Decay (CRAND) which is the product of cosmic ray interacting with the Earth's atmosphere [10]. Hence, they followed the GCR fluxes that are anti-correlated with the solar activity. Meanwhile, the solar activity had not much effect in the case of GCR and trapped electrons.

However, it did have an impact towards the NIEL values for GCR as shown in Figure 3. Their values were higher during solar minimum for most of the solar cycles except for solar cycle 23 that showed an entirely different result. It deviated from the rest by having a higher NIEL during solar maximum. Ordinarily, GCR fluxes, which contribute to the value level of LET and NIEL, are at the greatest amount during low solar activity due to less attenuation from the solar wind [1]. Therefore, one would expect the radiation doses to also be higher during solar minimum. However, their values in solar cycle 23 showed otherwise.

While we initially assumed that there might a technical error behind this but by looking into other monthly period of the same solar maximum within solar cycle 23, we found that most of the months also showed significant higher NIEL values than the one derived during solar minimum which was taken on March 2007 (refer Figure 4). An interesting aspect to solar cycle 23 is that its maximum period had double peaks, declining from the first peak before rising again into a second peak and later on progressing into solar minimum (http://nextgrandminimum.wordpress.com/2013/02/23/a-graphical-comparison-of-solar-cycles-21-22-23-and-24-2). Based on Figure 5, the intensity of GCR in solar cycle 23 remained to be low at high solar activity and increased at low solar activity. The pattern is the same for the other three solar cycles. Hence, the GCR flux was not the main factor behind the anomaly in its NIEL values here but from some other unforeseen parameters.

Although the LET of GCR is the lowest, however, in terms of NIEL, GCR gave the same level of damage as trapped protons do. This could be said that despite its low population within LEO/NEqO orbit, GCR also had the tendency to cause damage through non-ionizing mechanisms. The NIEL values for trapped electrons, however, were too low, by the factor of 200 lower in comparison with NIEL values of GCR and trapped protons and thus, are relatively insignificant. At the same time, no LET and NIEL values produced by SEP, suggesting that there were no SEP fluxes detected within the Razaksat altitude. This could mean that the particles were effectively shielded by the Earth's magnetosphere, particularly at the equatorial region so that none reached the low altitudes.

Table 2. LET by three radiation sources through GaAs planar geometry at 685km altitude following Razaksat orbital data during solar cycle 21 until December 2013 of solar cycle 24

	Radiation Sources	LET (MeV/cm) in 1mm GaAs planar	
		Solar Minimum	Solar Maximum
Solar Cycle 21	GCR	8.445 ± 0.778	8.479 ± 0.783
	SPE	0	0
	Trapped Proton	35.770 ± 21.860	27.810 ± 18.140
	Trapped Electron	10.840 ± 3.821	10.030 ± 3.432
Solar Cycle 22	GCR	8.552 ± 1.315	8.470 ± 0.773
	SPE	0	0
	Trapped Proton	36.090 ± 23.110	27.260 ± 16.030
	Trapped Electron	10.840 ± 3.821	10.030 ± 3.432
Solar Cycle 23	GCR	8.528 ± 1.056	8.408 ± 0.237
	SPE	0	0
	Trapped Proton	38.890 ± 23.050	29.470 ± 19.480
	Trapped Electron	10.840 ± 3.821	10.030 ± 3.431
Solar Cycle 24	GCR	8.469 ± 0.817	8.384 ± 0.250
	SPE	0	0
	Trapped Proton	39.050 ± 21.430	26.900 ± 18.010
	Trapped Electron	10.840 ± 3.821	10.030 ± 3.432

Table 3. NIEL by three radiation sources through GaAs planar geometry at 685km altitude following Razaksat orbital data during solar cycle 21 until December 2013 of solar cycle 24

	Radiation Sources	NIEL ($MeVcm^2/g$) in 1mm GaAs planar	
		Solar Minimum	Solar Maximum
Solar Cycle 21	GCR	$12.170 \times 10^{-3} \pm 47.460 \times 10^{-3}$	$7.256 \times 10^{-3} \pm 6.805 \times 10^{-3}$
	SPE	0	0
	Trapped Proton	$6.926 \times 10^{-3} \pm 4.051 \times 10^{-3}$	$6.594 \times 10^{-3} \pm 4.974 \times 10^{-3}$
	Trapped Electron	$3.012 \times 10^{-5} \pm 3.466 \times 10^{-5}$	$1.195 \times 10^{-5} \pm 1.197 \times 10^{-5}$
Solar Cycle 22	GCR	$7.652 \times 10^{-3} \pm 11.430 \times 10^{-3}$	$7.065 \times 10^{-3} \pm 9.581 \times 10^{-3}$
	SPE	0	0
	Trapped Proton	$5.890 \times 10^{-3} \pm 2.516 \times 10^{-3}$	$6.208 \times 10^{-3} \pm 3.601 \times 10^{-3}$
	Trapped Electron	$3.012 \times 10^{-5} \pm 3.466 \times 10^{-5}$	$1.195 \times 10^{-5} \pm 1.197 \times 10^{-5}$
Solar Cycle 23	GCR	$8.090 \times 10^{-3} \pm 11.500 \times 10^{-3}$	$11.270 \times 10^{-3} \pm 36.630 \times 10^{-3}$
	SPE	0	0
	Trapped Proton	$8.148 \times 10^{-3} \pm 7.142 \times 10^{-3}$	$6.527 \times 10^{-3} \pm 3.552 \times 10^{-3}$
	Trapped Electron	$3.012 \times 10^{-5} \pm 3.466 \times 10^{-5}$	$1.195 \times 10^{-5} \pm 1.197 \times 10^{-5}$
Solar Cycle 24	GCR	$9.369 \times 10^{-3} \pm 22.830 \times 10^{-3}$	$9.151 \times 10^{-3} \pm 24.980 \times 10^{-3}$
	SPE	0	0
	Trapped Proton	$7.271 \times 10^{-3} \pm 5.498 \times 10^{-3}$	$7.506 \times 10^{-3} \pm 8.540 \times 10^{-3}$
	Trapped Electron	$3.012 \times 10^{-5} \pm 3.466 \times 10^{-5}$	$1.195 \times 10^{-5} \pm 1.197 \times 10^{-5}$

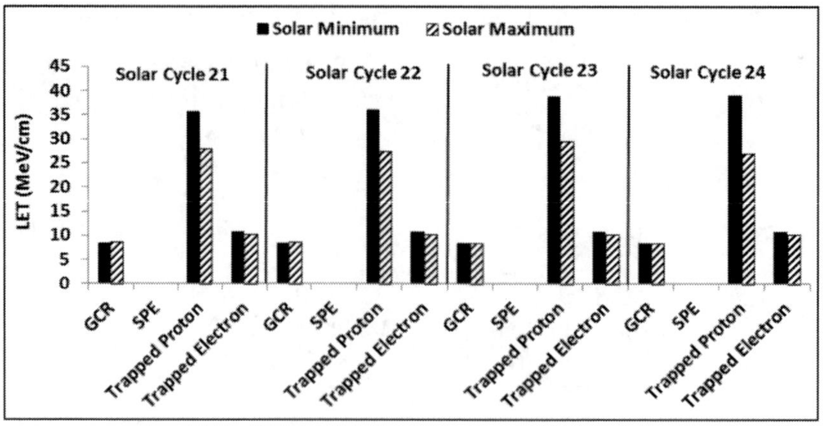

Figure 2. LET values by three radiation sources in LEO/NEqO orbit through a 1mm GaAs planar during selected solar minimum and solar maximum for solar cycle 21 until December 2013 of solar cycle 24

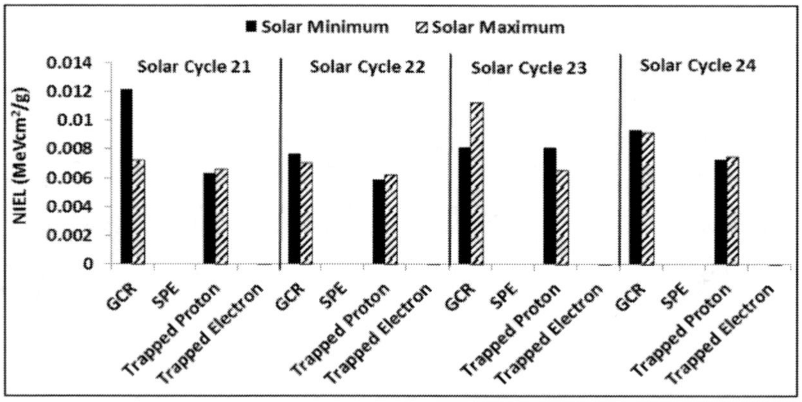

Figure 3. NIEL values by three radiation sources in LEO/NEqO orbit through a 1mm GaAs planar during selected solar minimum and solar maximum for solar cycle 21 until December 2013 of solar cycle 24

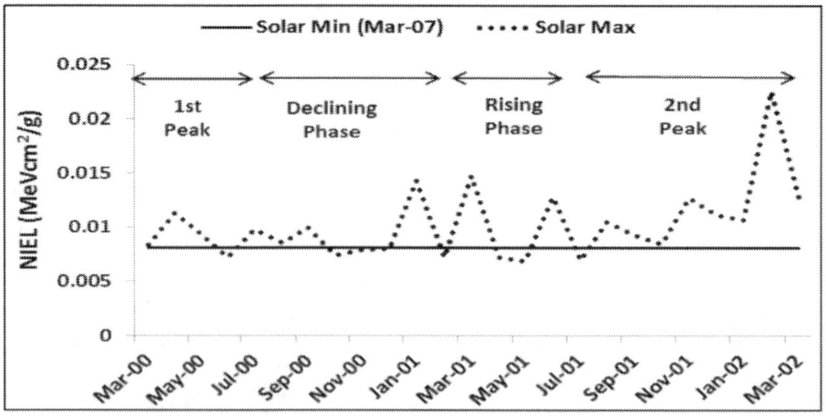

Figure 4. NIEL values by GCR during double peaks solar maximum in solar cycle 23 where most are above the NIEL value of solar minimum March 2007

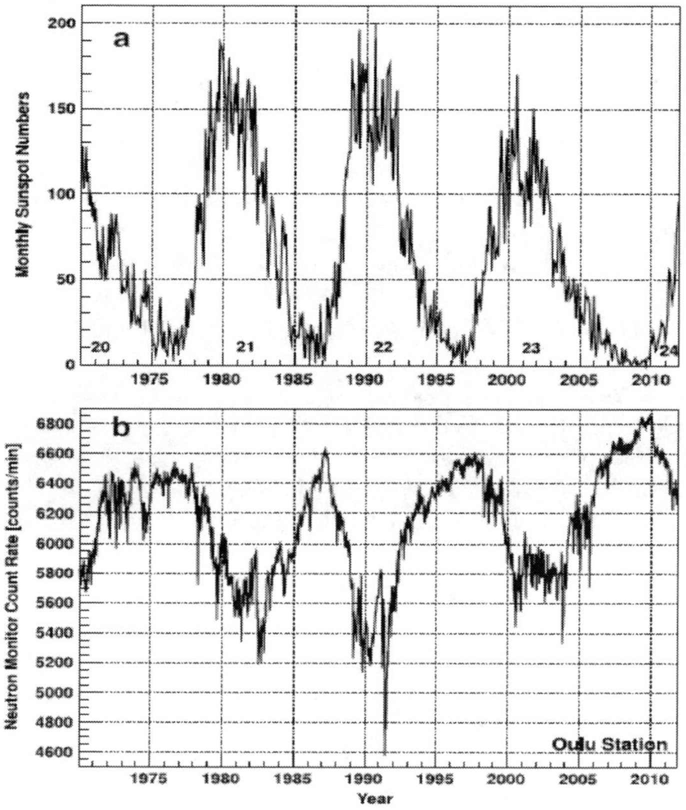

Figure 5. Monthly sunspot numbers for solar cycles 20 to 24 from 1970 until end of 2011 (a) corresponding to the 10-day averaged neutron monitor count rates from Oulu Station (b) (image taken from [11])

4. Conclusion and Future Work

Trapped protons gave the highest LET during solar minimum, followed by GCR and trapped electrons. The LET of trapped protons was two times higher than trapped electrons due to its dominant population in the inner radiation belt and also in SAA. GCR gave the same level of displacement damage towards GaAs as trapped protons based on their NIEL values. The NIEL value of GCR was unusually high during solar maximum in solar cycle 23 indicating abnormality for non-ionizing mechanisms occurring only during solar cycle 23 at its maximum period while the other three solar cycles showed similar pattern. We also identified that flux was not the main cause behind the deviated value but from other unforeseen parameters. The NIEL values for trapped electrons were relatively insignificant by a factor of 200 with that of trapped protons. At 685km altitude and 9° inclination, no SEP was detected to give LET and NIEL values due to the effective shielding of the Earth's magnetic field.

The abnormality of non-ionizing mechanisms by GCR during high solar activity in the 23rd cycle required further study by considering the application of other models such as CRÈME96 and to do comparison between the models so the factor behind the abnormality can be determined. Also, it would be interesting to look at the same mechanism in other materials such as silicon carbon (SiC) and indium phosphorus (InP) so more information regarding the abnormality and the effects of the space energetic particles can be acquired.

Acknowledgements

This research is supported by Ministry of Education Malaysia (MoE) under grant FRGS/2/2013/SG03/UKM/02/3. We would also like to thank the SPENVIS team for providing the software tool.

Journal of Physics: Conference Series **539** (2014) 012010 doi:10.1088/1742-6596/539/1/012010

References

[1] Benton E R and Benton E V 2001 *Nucl. Instr. and Meth. In Phys. Res. B* **184** 255

[2] Suparta W and Zulkeple S K 2014 *paper presented at the 2014 International Conference on Science and Engineering in Mathematics, Chemistry and Physics* Jakarta 13-14 January 2014

[3] Dyer C 2001 www.esa-spaceweather.net/spweather/workshops/SPW_W3/PROCEEDINGS_W3/solspa1.pdf

[4] Bourdarie S and Xapsos M 2008 *IEEE Trans. in Nucl. Sc. (TNS)*

[5] Jun I, Xapsos M A, Messenger S R, Burke E A, Walters R J, Summers G P and Jordan T 2003 *IEEE Trans. Nucl. Sci.* **50** 1924

[6] Kruglanski M, Messios N, Donder E D, Gamby E, Calders S, Hetey L, Evans H and Daly E 2009 *IEEE* 978-1-4577-0493-2

[7] International Standard ISO 15390 2004

[8] Feynman J, Spitale G, Wang J and Gabriel S 1993 *J. Geophys. Res.* **98** 13281

[9] Vette J I 1991 *NSSDC/WDC-A-R&S 91-29*

[10] Albert J M, Ginet G P and Gussenhoven M S 1998 *J. Geophys. Res.* **103** 9261

[11] Mrigakshi A I, Matthia D, Berger T, Reitz G and Wimmer-Schweingruher R F 2013 *Adv. Space. Res.* **52** 979

Holographic Screening Length on Parallel Motion of Quark-Antiquark Pair in Four Dimensional Strongly Coupled $\mathcal{N} = 4$ super-Yang-Mills plasma

ARDIAN NATA ATMAJA
Theoretical and Computational Group-Physics Research Center
Indonesian Institute of Science, Kompleks PUSPIPTEK Serpong
Tangerang Selatan, Banten 15314, Indonesia

E-mail: ardian@teori.fisika.lipi.go.id

Abstract. We study the screening length of a quark-antiquark pair moving in a strongly coupled hot plasma of $\mathcal{N} = 4$ super-Yang-Mills using AdS/CFT correspondence where the background metric is five dimensional AdS black hole. We take the string solution as such the separation length L of quark-antiquark pair is parallel to the string velocity v. The screening length and the bound energy are computed numerically using Mathematica. We find that the plots are bounded from below by some functions that are related to the momentum flow of the drag force configuration P_c. We compare the result by computing the screening length in the quark-antiquark reference frame by boosting the AdS black hole.

1. Introduction

The existence of Quark Gluon Plasma produced in heavy ion collision's experiment at RHIC and the new LHC can be indicated by the suppression of J/ψ meson production, $c\bar{c}$ pair. This phenomena is understood qualitatively when temperature of the plasma is larger than the Hagedorn temperature where the potential interaction of $c\bar{c}$ pair would not able to hold them anymore so that J/ψ meson will dissociate and be screened in the Quark Gluon Plasma[1]. The screening potential of $c\bar{c}$ pair depends on a separation length between c and $\bar{c}$, L which could have a value up to some maximum length L_{max}, called screening length, where beyond this length the screening potential becomes flat.

From the string theory point of view, a heavy quarks pair described by a Wilson loop is defined as an open string where both ends attached on the probe brane. Evaluating the Wilson loop will tell us information about the screening length dependent of the quark-antiquark potential. The procedure to evaluate the Wilson loop goes by extremizing the action of corresponding gravity dual theory [2, 3, 4]. A rigorous calculation on the screening length of a moving quark-antiquark pair at finite temperature four dimensional $\mathcal{N} = 4$ Supersymmetric Yang-Mills theory from AdS/CFT correspondence was pioneered by Liu, Rajagopal, and Wiedemann [5]. It was then generalized to arbitrary dimension for conformal and non-conformal gauge theories by Cáceres, Natsuume, and Okamura [6]. The calculations there were done by going to reference frame of the moving quark-antiquark pair or explicitly by boosting the background metric to the direction of quark-antiquark's velocity. A different approach was done in the reference frame of the plasma by Chernicoff, Garcia, and Guijosa [7] and furthermore they also calculated the energy from both approaches. One interesting result of these calculations is the present of boost factor, $(1 - v^2)$, in the screening length. In Ref. [6], it was shown that the scale of this factor depends on the dimension of black hole backgrounds. However, this scaling factor is valid in ultra-relativistic limit and disagrees with numerical fitting that was found in Ref. [7].

In this article, we are interested in computing the screening length of a heavy quark-antiquark, $q\bar{q}$, pair moving in the strongly coupled $\mathcal{N} = 4$ super-Yang-Mills plasma where the corresponding background metric is five dimensional AdS black hole in Poincare coordinate. Here, we consider the separation length to be parallel with the velocity. We further would like to make comparison with the computation in the reference frame of the $q\bar{q}$ pair where the plasma is moving hence the background metric is a boosting black hole which can be obtained simply by boosting procedure as in Ref. [6].

2. Screening Length

For our purpose here the corresponding background metric is the five dimensional AdS black hole [8]

$$\mathrm{d}s^2 = \frac{r^2}{R^2}\left(-h(r)\mathrm{d}t^2 + \mathrm{d}\vec{x}^2\right) + \frac{R^2}{r^2}\frac{1}{h(r)}\mathrm{d}r^2, \tag{1}$$

$$h(r) = 1 - \frac{r_H^4}{r^4}, \qquad T_H = \frac{r_H}{\pi R^2}, \tag{2}$$

where R is the radius of curvature of AdS space, T_H is the Hawking temperature, and r_H is the even horizon. We ignore the compact part of the space, S^5, as it can be effectively removed in this case. This background metric in AdS/CFT dictionary corresponds to the strongly coupled $\mathcal{N} = 4$ super-Yang-Mills through the following relation [10]

$$\frac{R^4}{\alpha'^2} = \lambda = g_{YM}^2 N_c, \qquad \lambda \gg 1, \tag{3}$$

where $\sqrt{\alpha'}$ is the fundamental string length scale, λ is the 't Hooft coupling, g_{YM} is the Yang-Mills coupling, and N_c is the number of color in $SU(N_c)$ gauge symmetry.

The dynamic of the string is classically given by the Nambu-Goto action under the background (1),

$$S = -\frac{1}{2\pi\alpha'}\int \mathrm{d}\sigma^2\sqrt{-\det g_{\alpha\beta}}, \qquad g_{\alpha\beta} \equiv G_{\mu\nu}\partial_\alpha X^\mu \partial_\beta X^\nu, \tag{4}$$

where σ^α is the worldsheet coordinate, with $\sigma^\alpha \equiv (\tau, \sigma)$, $X^\mu(\sigma^\alpha)$ is the space-time coordinate where the string worldsheet is embedded, and $G_{\mu\nu}$ is the space-time metric (1). From there, we obtain the equation of motion [9]

$$\nabla_\alpha P_\mu^\alpha = 0,$$

$$P_\mu^\alpha = \frac{\pi_\mu^\alpha}{\sqrt{-g}}, \qquad \pi_\mu^\alpha = \frac{\delta S}{\delta \partial_\alpha X^\mu}, \tag{5}$$

where π_μ^α is the cannonical worldsheet momentum with $g = \det g_{\alpha\beta}$.

Using the usual gauge $\tau = t$, while for σ will be determined later for convenient, we choose the string to be moving in the direction of x_3−axis with the following ansatz

$$x_3(\sigma^\alpha) \equiv v\tau + x_3(\sigma), \tag{6}$$

$$r(\sigma^\alpha) \equiv r(\sigma), \tag{7}$$

thus the equations of motion are given by

$$x_3'^2\left(r^4-r_H^4\right)'+r'^2\left(\frac{r^4(1-v^2)-r_H^4}{r^4-r_H^4}\right)'-2r'\sqrt{-g}\frac{\partial}{\partial\sigma}\left[\frac{r'}{\sqrt{-g}}\frac{r^4(1-v^2)-r_H^4}{r^4-r_H^4}\right]=0, \quad (8)$$

$$\partial_\sigma\pi_{x_3}^\sigma\equiv\frac{\partial}{\partial\sigma}\left[\frac{r^4-r_H^4}{\sqrt{-g}}x_3'\right]=0, \quad (9)$$

with $\pi_{x_3}^\sigma$ is constant and $'\equiv\frac{\partial}{\partial\sigma}$. For quark-antiquark system, the solution for r must satisfy conditions

$$r(-L/2)=r(L/2)=+\infty, \qquad r'(\sigma_p)=0, \quad (10)$$

where the length of string is L and $-L/2<\sigma_p<L/2$ is the turning point where r takes the minimum value $r(\sigma_p)=r_p\geq r_H$. Using the symmetry of the ends of the string, we can set $\sigma_p=0$. Another gauge that we might use is $x_3(\sigma)=\sigma$ where both equations of motion are similar. We can simply choose a gauge $x_3=v\tau+\sigma$ in the first place which gives us the equation (8) that has more complicated form than equation (9).

As both equations are similar we prefer to use equation (9) which is simpler to find the separation length, now is precisely L, of quark-antiquark pair. The solutions for r are given by

$$r'=\pm\frac{r^4-r_H^4}{P}\sqrt{\frac{r^4-(P^2+r_H^4)}{r^4(1-v^2)-r_H^4}}, \quad (11)$$

where $P\equiv\pi_{x_3}^\sigma$ is a constant denoting amount of x_3-component of momentum flow or transfer on the string. The force on the components of string must correspond to the direction of string motion which is in x_3-coordinate thus we interpret the force proportional to $\pi_{x_3}^\sigma$ instead of π_r^σ. For quark-antiquark pair $P>0$, where $P=0$ would correspond to a straight string configuration, and the positive sign in (11) is proportional to the energy flows from the boundary down to $r=r_p$ opposed to the negative sign. The quark-antiquark pair can be built out of two configurations (a) and (b) as shown in Figure 1 with

$$\frac{\mathrm{d}r}{\mathrm{d}\sigma}=\begin{cases}\frac{r^4-r_H^4}{P}\sqrt{\frac{r^4-(P^2+r_H^4)}{r^4(1-v^2)-r_H^4}} & ,0<\sigma\leq L/2\\ 0 & ,\sigma=0\\ -\frac{r^4-r_H^4}{P}\sqrt{\frac{r^4-(P^2+r_H^4)}{r^4(1-v^2)-r_H^4}} & ,-L/2\leq\sigma<0\end{cases}. \quad (12)$$

Both solutions (11) only different by a sign $P\to-P$ therefore we might want to solve only one of them. At $r(0)=r_p$, we must have $r_p^4-(P^2+r_H^4)=0$ which implies that $r_p>r_H$ for $P>0$. There is a critical radius $r_c^4=\frac{r_H^4}{1-v^2}$ and it is also $r_c>r_H$ for $v\neq0$. If $r_p>r_c$ then the geometry is cut-off below at $r=r_p$ and string configuration for the quark-antiquark bound is formed. If $r_p<r_c$ the geometry is also cut-off at $r=r_c$ however this solution is not physical because the string is singular in the middle at $r=r_c$ so we will ignore this solution. If $r_p=r_c$ then the condition (10) could not be satisfied and the string configuration physically would be similar to the drag force configuration thus for quark-antiquark pair we must have $r_p>r_c$ or $P^2>v^2r_c^4$. So for some fixed velocity v, or r_c, the string configuration is characterized by amount of momentum transfer P on the string with $P^2\geq v^2r_c^4$. At $P^2=P_c^2=v^2r_c^4$, the string tends to represent a single quark and if we increase the momentum transfer P the quark-antiquark pair is formed.

From equation (11), we obtain an integral formula for computing separation length as below

$$\int_0^{L/2}\mathrm{d}\sigma=\frac{L}{2}=\int_{r_p}^{\infty}\mathrm{d}r\frac{P}{r^4-r_H^4}\sqrt{\frac{r^4(1-v^2)-r_H^4}{r^4-(P^2+r_H^4)}}. \quad (13)$$

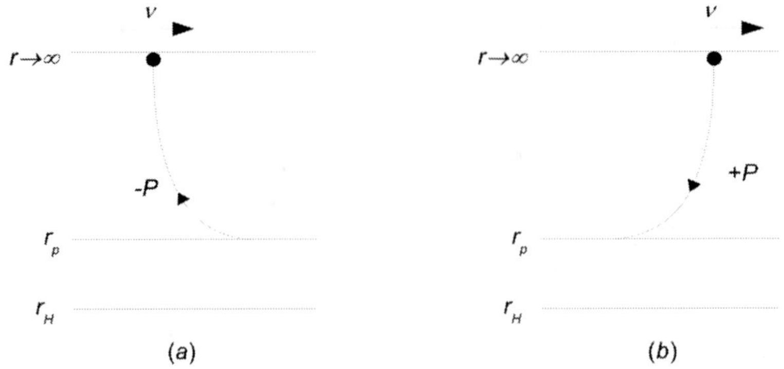

Figure 1: The positive sign solution (b) joints up with the negative sign solution (a) resulting in the quark-antiquark pair.

We cannot solve the integral above analytically so we are going to plot the integral numerically. The numerical solution of the separation length for various value of v is shown in Figure 2. Our calculation here is in the same foot as in Ref. [6] for parallel case, $\theta = 0$, where they only discuss the ultra-relativistic limit and analytically compute the scaling factor of the screening length.

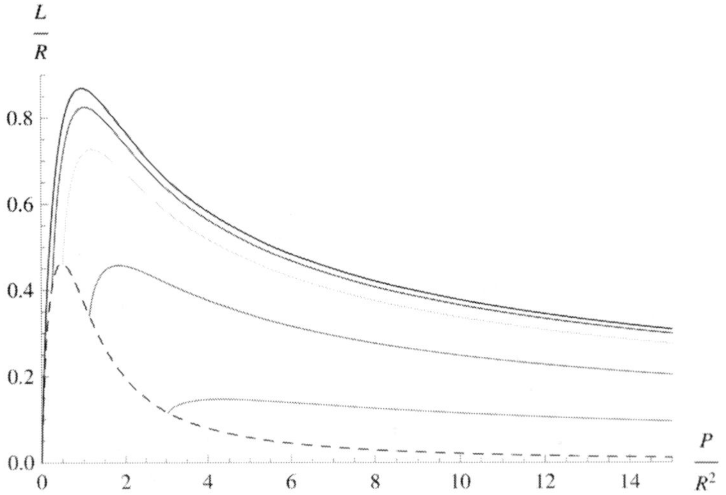

Figure 2: Plot of dimensionless separation length, L/R, for various $v = 0, 0.25, 0.45, 0.75, 0.95$, with fixed $r_H/R = 1$, as function of P/R^2 where the corresponding colors are black, blue, green, brown, and red. The dashed black line is the saturated separation length obtained by substituting $(1 - v^2) = r_H^4/r_p^4$, for which $r_c = r_p$, into formula (13). In this line the quark-antiquark pair can not be formed instead the string becomes a single quark configuration.

As one can see, we produce similar profile of screening length L_s as in the perpendicular case, Ref. [7], except that the momentum transfer is bounded from below at $P^2 = P_c^2 \equiv v^2 r_c^4$, for fixed v, where below this value the integral (13) becomes complex. Therefore our plot do not start from $P/R^2 = 0$, which is different from the plot produced in Ref. [7], instead we start from $P = P_c$. Similar to the perpendicular case, the separation length in Figure 2 consist of two regions separated by the maximum L_{max} called the screening length at the point $P = P_{max}$. For

Journal of Physics: Conference Series **539** (2014) 012011 doi:10.1088/1742-6596/539/1/012011

$P_c < P < P_{max}$, the string configuration is metastable while for $P \geq P_{max}$ is stable, for detail discussion see Ref. [7].

Unlike the perpendicular case, this quark-antiquark pair of string configuration requires an external force to keep it moves with a constant velocity. The amount of required external force is proportional to P_c suggested from the drag force computation[10]. The momentum flow P on the string is bigger then the external force and so it is natural to think that the difference is related to the potential to keep quark-antiquark pair bounded. In this sense, the quak-antiquark pair still feels a drag force by the present of this external force. One can see from the Figure 2 the larger momentum flow P, for $P > P_{max}$, the smaller separation length thus one can consider the quark-antiquark pair effectively as a particle. In addition, since the tip of the string at r_p is farther from the horizon and is closer to the boundary, $r_p \to \infty$, the quark-antiquark pair is insensitive to the presence of the plasma. The string has a very short length in a very large P hence it might correspond to light quark-antiquark pair. This light quark-antiquark pair could be the constituent of other plasma located at the UV-brane, where the geometry is cut-off by the presence of UV-brane in order to remove the UV divergence and to make the quark mass, or correspondingly the string length, to be large but finite.

3. Total energy of the string

We are now interested to compute the total energy of the string. Energy density for the string in quark-antiquark pair setup above is given by [10]

$$\mathcal{H}_{bound} = 2T_0 \frac{(r^4 - r_H^4)^2 - r^4 P^2 v^2}{(r^4 - r_H^4)\sqrt{(r^4(1-v^2) - r_H^4)(r^4 - r_H^4 - P^2)}}, \qquad P^2 > v^2 r_c^4. \tag{14}$$

As usual the bound energy density is divergent at the boundary, $\sigma = \pm L/2$, therefore it must be substracted with the unbound energy density of two quark and antiquark moving with constant velocity v where the energy density is

$$\mathcal{H}_{unbound} = 2T_0 \frac{(r^4 - r_H^4)^2 - r^4 P_c^2 v^2}{(r^4 - r_H^4)\sqrt{(r^4(1-v^2) - r_H^4)(r^4 - r_H^4 - P_c^2)}}, \qquad P_c^2 = v^2 r_c^4. \tag{15}$$

This unbound energy density will cancel the boundary divergence of the bound energy density with the cost of a new divergence at the horizon. As explained in Ref. [7], and also Ref. [10], this source of divergence coming from the unbound energy density because of infinite amount of energy flowing down from boundary, supplied by external force, to the horizon to keep the quark and antiquark move with constant velocity. Furthermore this also produce some ambiguity in removing this horizon divergence.

As resolution we need to compute the energy density in the rest frame of the moving string and this can be done by the following coordinate transformation [11]:

$$t \to \gamma\,(t - v x_3)\,, \qquad x_3 \to \gamma\,(x_3 - vt)\,, \tag{16}$$

with $\gamma = \left(1 - v^2\right)^{-1/2}$ is the boost factor. The resulting five dimensional "Boost-AdS" black hole is

$$\mathrm{d}s^2 = -\frac{r^4 - r_H^4}{r^2 R^2}\gamma^2\,(\mathrm{d}t - v\mathrm{d}x_3)^2 + \frac{r^2 R^2}{r^4 - r_H^4}\mathrm{d}r^2 + \frac{r^2}{R^2}\gamma^2\,(\mathrm{d}x_3 - v\mathrm{d}t)^2 + \frac{r^2}{R^2}(\mathrm{d}x_1^2 + \mathrm{d}x_2^2). \tag{17}$$

Taking the same gauge, $t = \tau$ and $x_3 = \sigma$, the static solution for radial coordinates, $r \equiv r(\sigma)$, is given by

$$r' = \frac{r^4 - r_H^4}{P\gamma}\sqrt{\frac{r^4 - (P^2 + r_H^4)}{r^4\gamma^{-2} - r_H^4}}, \tag{18}$$

Journal of Physics: Conference Series **539** (2014) 012011 doi:10.1088/1742-6596/539/1/012011

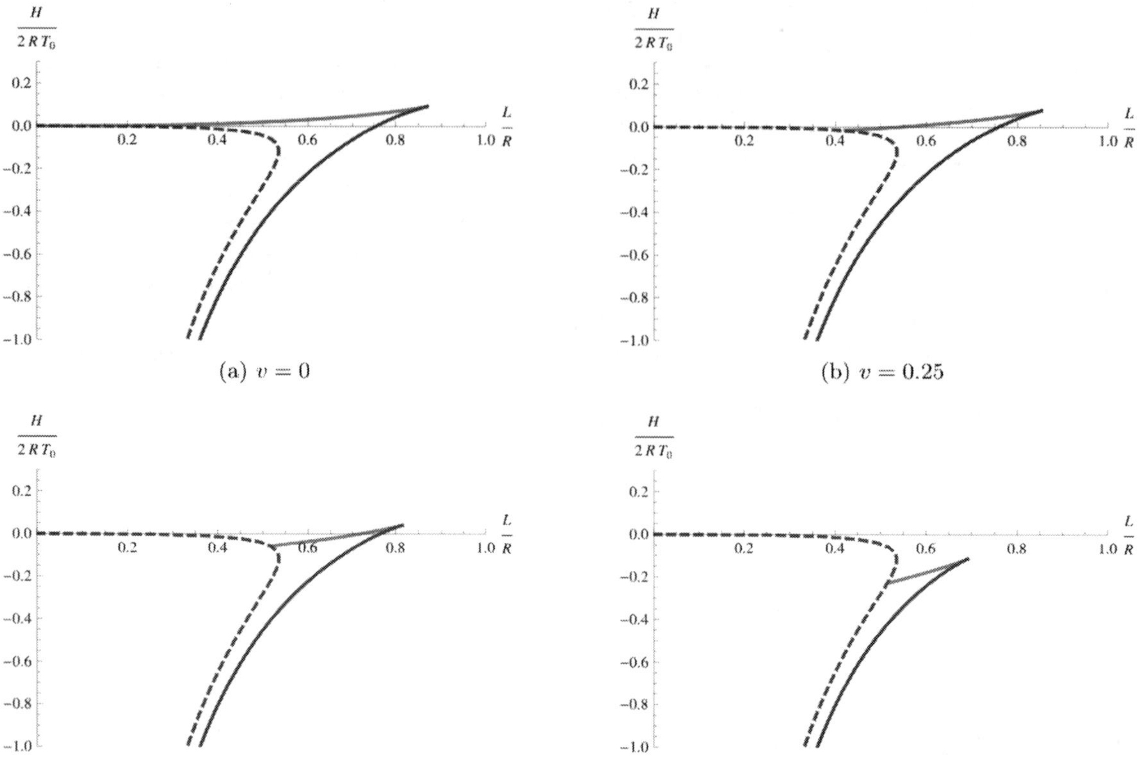

(a) $v = 0$ (b) $v = 0.25$

(c) $v = 0.45$ (d) $v = 0.75$

Figure 3: Plots of normalized energy for quark-antiquark pair $H/(2T_0R)$ as function of separation length L/R for various velocity v with fixed $r_H/R = 1$. The metastable configuration is colored with red while the stable configuration is colored with blue. The dashed black line is the saturated energy H_{sat} where all possible energies in this line are forbidden for the quark-antiquark pair configuration.

where P is again the same x_3-component of momentum transfer as in the corresponding static black hole case. One can immediately see from (18) that separation length will be different from the static case by a boost factor γ thus the screening length as well, $L_{Boost} = \gamma L_{AdS}$. In this "Boost-AdS" black hole the bound energy density turns out to be

$$\mathcal{H}_{bound} = 2T_0\sqrt{\frac{r^4 - r_H^4\gamma^2}{r^4 - r_H^4 - P^2}}, \qquad P^2 > v^2 r_c^4. \tag{19}$$

with the integral range $r_p \leq r < \infty$. The corresponding unbound energy density is simply written as

$$\mathcal{H}_{unbound} = 2T_0\sqrt{\frac{r^4 - r_H^4\gamma^2}{r^4 - r_H^4 - P_c^2}}, \qquad P_c^2 = v^2 r_c^4. \tag{20}$$

As we can see the boundary divergence of the bound energy is removed by the unbound energy without introducing the horizon divergence. So, we can define a normalized energy for quark-antiquark pair

$$H = \int_{r_p}^{\infty} \mathrm{d}r\ \mathcal{H}_{bound} - \int_{r_H}^{\infty} \mathrm{d}r\ \mathcal{H}_{unbound}. \tag{21}$$

Rewriting the normalized energy H as a function of separation length L instead of P using formula (13), we plot numerically the normalized energy for various value of velocity v, see Figure 3. The characteristic behaviour of the normalized energy is similar as in Ref. [7] except that the normalized energy for fixed velocity v has a minimum length in the metastable configuration, red line of Figure 3, given by the dashed black line. This dashed black line is produced by the saturated energy of the following function

$$H_{sat}(P) = -2T_0\left(r_p - r_H\right). \tag{22}$$

4. Conclusion and Discussion

In section 2, we have computed the separation length as a function of momentum transfer P and plotted it for various velocity v. In the plot of Figure 2, for fixed velocity v, the separation length has minimum momentum transfer $P_c > 0$ for $v \neq 0$. This is because P is bounded from below where its value has to be bigger than the drag force momentum transfer, $P_c = vr_c^2$. Below this value there is no physical solution for string configuration of quark-antiquark pair. Therefore right after the momentum transfer P passes this bound the quar-antiquark pair is formed at the boundary with non-zero separation length but this quark-antiquark pair is unstable or at least metastable. The reason is because the tip of the string in this configuration, located at r_p, is very close to r_c where at $v = 0$, $r_c = r_H$. However, there is another configuration of string, which is much more stable then the previous one, with the same separation length at high momentum transfer P where the tip of string, r_p, is far away from r_c. Both configurations separated by the maximum separation length called the screening length as discussed lengthy in Ref. [7]. An interesting line shown in Figure 2 is the line of bounded momentum transfer $P = P_c$ colored with black dashed. This line's profile is mimicking the lines's profile of the separation length, especially for $v = 0$, where there is a maximum length. We also computed the bound energy density of the quark-antiquark pair. Similarly the normalized bound energy of quark-antiquark pair also has minimum length on its metastable configuration(red) as shown by dashed line in Figure 3.

Acknowledgments

This work is supported by Kompetitif-LIPI 2013 grant with project's title: "Dinamika Fluida pada Energi Tinggi".

References

[1] Matsui T and Satz H 1986 *Phys. Lett. B* **178** 416
[2] Maldacena J M 1998 *Phys. Rev. Lett.* **80** 4859
[3] Rey S J and Yee J T 2001 *Eur. Phys. J. C* **22** 379
[4] Rey S J, Theisen S and Yee J T 1998 *Nucl. Phys. B* **527** 171
[5] Liu H, Rajagopal K and Wiedemann U A 2007 *Phys. Rev. Lett.* **98** 182301
[6] Caceres E, Natsuume M and Okamura T 2006 *JHEP* **0610** 011
[7] Chernicoff M, Garcia J A and Guijosa A 2006 *JHEP* **0609** 068
[8] Witten E 1998 *Adv. Theor. Math. Phys.* **2** 505
[9] Gubser S S 2006 *Phys. Rev. D* **74** 126005
[10] Herzog C P, Karch A, Kovtun P, Kozcaz C and Yaffe L G 2006 *JHEP* **0607** 013
[11] Bhattacharyya S, Hubeny V E, Minwalla S and Rangamani M 2008 *JHEP* **0802** 045

CTPNP 2014: "From Universe to String's Scale" IOP Publishing
Journal of Physics: Conference Series **539** (2014) 012012 doi:10.1088/1742-6596/539/1/012012

Nonzero θ_{13} and CP violation from broken $\mu-\tau$ symmetry

Asan Damanik

Faculty of Science and Technology, Sanata Dharma University, Kampus III USD, Paingan, Maguwoharjo, Sleman, Yogyakarta, Indonesia

E-mail: d.asan@lycos.com

Abstract. Nonzero and relatively large of θ_{13} mixing angle has some phenomenological consequences on neutrino physics beyond the standard model. One of the consequences if the mixing angle $\theta_{13}=0$ is the possibility of the CP violation on the neutrino sector. In order to obtain nonzero θ_{13} mixing angle, we break the neutrino mass matrix that obey $\mu-\tau$ symmetry by introducing a complex parameter and determine the Jarlskog invariant as a measure of CP violation existence. By using the experimental data θ_{12} and θ_{23} as input, we can determine the Dirac phase δ as function of mixing angle θ_{13}.

1. Introduction

Recently, there are two unsolved major problems related to neutrino physics i.e. underlying symmetry of neutrino mass matrix and neutrino mixing matrix when confronted to the experimental results. One of the interesting and popular underlying symmetry for neutrino mass matrix is the $\mu-\tau$ symmetry. The $\mu-\tau$ symmetry reduces the number of parameters in symmetric neutrino mass matrix from 6 parameters to 4 parameters. The neutrino mass matrix with $\mu-\tau$ symmetry can also obtained from three well-known neutrino mixing matrices i.e. tribimaximal (TBM), bimaximal (BM), and democratic (DC). But, the three well-known of neutrino mixing matrices predict mixing angle $\theta_{13}=0$ and hence Dirac phase δ can be arbitrary. The standard parametrization of neutrino mixing matrix (V) is given by:

$$V=\begin{pmatrix} c_{12}c_{13} & s_{12}c_{13} & s_{13}e^{-i\delta} \\ -s_{12}c_{23}-c_{12}s_{23}s_{13}e^{i\delta} & c_{12}c_{23}-s_{12}s_{23}s_{13}e^{i\delta} & s_{23}c_{13} \\ s_{12}s_{23}-c_{12}c_{23}s_{13}e^{i\delta} & -c_{12}s_{23}-s_{12}c_{23}s_{13}e^{i\delta} & c_{23}c_{13} \end{pmatrix} \quad (1)$$

where c_{ij} is the $\cos\theta_{ij}$, s_{ij} is the $\sin\theta_{ij}$, and θ_{ij} are the mixing angles. In the basis where the charged lepton mass matrix is diagonal, the neutrino mass matrix M_ν can be diagonalized by mixing matrix V as follow:

$$M_\nu=VMV^T, \quad (2)$$

Journal of Physics: Conference Series **539** (2014) 012012 doi:10.1088/1742-6596/539/1/012012

where the diagonal neutrino mass matrix M is given by:

$$M = \begin{pmatrix} m_1 & 0 & 0 \\ 0 & m_2 & 0 \\ 0 & 0 & m_3 \end{pmatrix}. \tag{3}$$

If we put the mixing angle $\theta_{13} = 0$ and consequently $s_{13} = 0$ and $c_{13} = 1$ in Eq. (1), then the neutrino mass matrix M_ν in Eq. (2) read:

$$M_\nu = \begin{pmatrix} P & Q & Q \\ Q & R & S \\ Q & S & R \end{pmatrix}, \tag{4}$$

where:

$$P = m_1 c_{12}^2 + m_2 s_{12}^2, \tag{5}$$
$$Q = (m_2 - m_1) c_{12} s_{12} c_{23}, \tag{6}$$
$$R = (m_1 s_{12}^2 + m_2 c_{12}^2) c_{23}^2 + m_3 s_{23}^2, \tag{7}$$
$$S = (-m_1 s_{12}^2 - m_2 c_{12}^2 + m_3) s_{23} c_{23}. \tag{8}$$

It is apparent from neutrino mass matrix (M_ν) in Eq. (4) that the neutrino mass matrix deduced from neutrino mixing matrix with assumption $\theta_{13} = 0$ is $\mu - \tau$ symmetry. But, as dictated from the experimental results, the mixing angle $\theta_{13} \neq 0$ and relatively large [1, 2, 3, 4, 5] which imply that the assumption $\theta_{13} = 0$ in formulating the neutrino mixing matrix must be rule out and hence the exact $\mu - \tau$ symmetry as the underlying symmetry of neutrino mass matrix is no longer adequate to accommodate the recent experimental results.

In order to accommodate the nonzero θ_{13} and Jarlskog rephasing invariant $J_{\rm CP}$ as a measure of CP violation in neutrino sector, in this paper we break the $\mu - \tau$ symmetry softly by introducing a one small parameter to break the neutrino mass matrix with $\mu - \tau$ symmetry softly. The paper is organized as follow: in section 2 we break the neutrino mass matrix with $\mu - \tau$ symmetry by introducing a complex parameter into neutrino mass matrix and in section 3 we determine $J_{\rm CP}$ as function of the mixing angle θ_{13} by using the experimental results of mixing angles θ_{12} and θ_{23} as input. Finally, section 4 is devoted to conclusions.

2. Broken $\mu - \tau$ symmetry

Concerning the neutrino mass matrix that obey the $\mu - \tau$ symmetry and mixing angle θ_{13}, Mohapatra [6] stated explicitly that neutrino mass matrix that obey $\mu - \tau$ symmetry to be the reason for maximal $\mu - \tau$ mixing and one gets $\theta_{13} = 0$, conversely if $\theta_{13} \neq 0$ can provide the $\mu - \tau$ symmetry beraking manifests in the case of normal hierarchy. Aizawa and Yasue [7] analysis complex neutrino mass texture and the $\mu - \tau$ symmetry which can yield small θ_{13} as a $\mu - \tau$ breaking effect. The $\mu - \tau$ symmetry breaking effect in relation with the small θ_{13} also discussed in [8]. Analysis of the correlation between CP violation and the $\mu - \tau$ symmetry breaking can be read in [9, 10, 11, 12].

Now, we are in position to study the effect of neutrino mass matrix that obey the $\mu - \tau$ symmetry breaking in relation to the nonzero θ_{13} and Jarlskog rephasing invariant $J_{\rm CP}$ by breaking the neutrino mass matrix in Eq. (4). We break the neutrino mass matrix in Eq. (4) by introducing a complex paramater ix with the constraint that the trace of the broken neutrino mass matrix is remain constant or equal to the trace of the unbroken one. This scenario of breaking has been applied by Damanik [13] to break the neutrino mass matrix invariant under a cyclic permutation. In this breaking scenario, the broken neutrino mass matrix reads:

$$M_\nu = \begin{pmatrix} P & Q & Q \\ Q & R - ix & S \\ Q & S & R + ix \end{pmatrix}. \tag{9}$$

As stated previously that the CP violation can be determined from the Jarlskog rephasing invariant $J_{\rm CP}$. Alternatively, Jarlskog rephasing invariant $J_{\rm CP}$ can be determined using the relation [14]:

$$J_{\rm CP} = -\frac{\text{Im}\left[(M_\nu^{'})_{e\mu}(M_\nu^{'})_{\mu\tau}(M_\nu^{'})_{\tau e}\right]}{\Delta m_{21}^2 \Delta m_{32}^2 \Delta m_{31}^2}, \tag{10}$$

where $(M_\nu^{'})_{ij} = (M_\nu M_\nu^\dagger)_{ij}$ with $i, j = e, \nu, \tau$, and $\Delta m_{kl}^2 = m_k^2 - m_l^2$ with $k = 2, 3,$ and $l = 1, 2$. From Eq. (9), we have:

$$M_\nu^{'} = \begin{pmatrix} P^2 + 2Q^2 & Q(P + S + R + ix) & Q(P + R + S - ix) \\ Q(P + S + R - ix) & Q^2 + R^2 + S^2 + x^2 & Q^2 + 2S(R - ix) \\ Q(P + S + R + ix) & Q^2 + 2S(R + ix) & Q^2 + R^2 + S^2 + x^2 \end{pmatrix}. \tag{11}$$

3. Nonzero θ_{13} and Jarlskog rephasing invariant

From Eqs. (10) and (11) we have the Jarlskog rephasing invariant as follow:

$$J_{\rm CP} = \frac{2Q^2\left[S(S + P)^2 - Q^2(R + S + P) - R^2 S\right] x - (2Q^2 S)x^3}{\Delta m_{21}^2 \Delta m_{32}^2 \Delta m_{31}^2}. \tag{12}$$

If we insert Eqs. (5)-(8) into Eq. (12), then we have the $J_{\rm CP}$ as follow:

$$\begin{aligned} J_{\rm CP} = & \frac{2(m_2 - m_1)^2 c_{12}^2 s_{12}^2 c_{23}^2}{\Delta m_{21}^2 \Delta m_{32}^2 \Delta m_{31}^2}[[(-m_1 s_{12}^2 - m_2 c_{12}^2 + m_3)s_{23}c_{23} \\ & \times((-m_1 s_{12}^2 - m_2 c_{12}^2 + m_3)s_{23}c_{23} + m_1 c_{12}^2 + m_2 s_{12}^2)^2 \\ & -(m_2 - m_1)^2 c_{12}^2 s_{12}^2 c_{23}^2((m_1 s_{12}^2 + m_2 c_{12}^2)c_{23}^2 + m_2 s_{23}^2 \\ & -(m_1 s_{12}^2 + m_2 c_{12}^2 - m_3)s_{23}c_{23} + m_1 c_{12}^2 + m_2 s_{12}^2) \\ & -((m_1 s_{12}^2 + m_2 c_{12}^2)c_{23}^2 + m_2 s_{23}^2)^2(-m_1 s_{12}^2 - m_2 c_{12}^2 \\ & +m_3)s_{23}c_{23}]x - c_{23}s_{23}[-m_1 s_{12}^2 - m_2 c_{12}^2 + m_3]x^3]. \end{aligned} \tag{13}$$

From Eq. (13), one can see that the exact $\mu - \tau$ symmetry ($x = 0$) lead to $J_{\rm CP} = 0$ as claimed by many authors, and if we still want to have $J_{\rm CP} \neq 0$ from the $\mu - \tau$ symmetry, then we must break it softly.

From Eq. (13) one can see that in this breaking scenario, the Jarlskog rephasing invariant ($J_{\rm CP}$) in neutrino sector can be proceed if we break the neutrino mass matrix of $\mu - \tau$ symmetry by introducing complex parameter ix parameter. It is also apparent from Eq. (13) that $m_1 \neq m_2$ as a constraint to the existence of the $J_{\rm CP} \neq 0$.

In order to get the value of Jarlskog rephasing invariant $J_{\rm CP}$ of Eq. (13), we use the experimental values of neutrino oscillation as input. The current combined world data for neutrino squared-mass difference are given by [15, 16]:

$$\Delta m_{21}^2 = 7.59 \pm 0.20(^{+0.61}_{-0.69}) \times 10^{-5}\ \mathrm{eV}^2, \quad (14)$$
$$\Delta m_{32}^2 = 2.46 \pm 0.12(\pm 0.37) \times 10^{-3}\ \mathrm{eV}^2,\ (\text{for NH}) \quad (15)$$
$$\Delta m_{32}^2 = -2.36 \pm 0.11(\pm 0.37) \times 10^{-3}\ \mathrm{eV}^2,\ (\text{for IH}) \quad (16)$$
$$\theta_{12} = 34.5 \pm 1.0(^{3.2}_{-2.8})^o,\ \ \theta_{23} = 42.8^{+4.5}_{-2.9}(^{+10.7}_{-7.3})^o,\ \ \theta_{13} = 5.1^{+3.0}_{-3.3}(\leq 12.0)^o, \quad (17)$$

at 1σ (3σ) level. We also put $m_1 = 0$ as a first approximation and within this scenario, for normal hierarchy (NH):

$$m_2^2 = \Delta m_{21}^2, \quad (18)$$
$$m_3^2 = \Delta m_{32}^2 + \Delta m_{21}^2. \quad (19)$$

By inserting the values of Eqs. (14)-(19), we have:

$$J_{\mathrm{CP}} = 0.4644x - 832.9790x^3. \quad (20)$$

From Eq. (20), we can determine the maximum value of J_{CP} by using the relation:

$$\frac{dJ_{\mathrm{CP}}}{dx} = 0, \quad (21)$$

which proceed $x = 0.0167$. By substituting the value of $x = 0.0167$ into Eq. (20), we have the maximum value of Jarlskog rephasing invariant:

$$J_{\mathrm{CP}} \approx 0.004. \quad (22)$$

It is also possible to determine the Jarlskog rephasing invariant J_{CP} from the neutrino mixing matrix by using the relation:

$$J_{\mathrm{CP}} = \mathrm{Im}(V_{11}^* V_{23}^* V_{13} V_{21}). \quad (23)$$

From neutrino mixing matrix of Eq. (1), the Jarlskog rephasing invariant J_{CP} reads:

$$\begin{aligned} J_{\mathrm{CP}} &= c_{12}s_{12}c_{23}s_{23}c_{13}^2 s_{13} \sin\delta \\ &= c_{12}s_{12}c_{23}s_{23}\left(s_{13} - s_{13}^3\right)\sin\delta. \end{aligned} \quad (24)$$

If we put the experimental values of mixing angles θ_{12} and θ_{23} of Eq. (17) into Eq. (24), then we have:

$$J_{\mathrm{CP}} = 0.2327\left(s_{13} - s_{13}^3\right)\sin\delta. \quad (25)$$

Since the value of $s_{13}^3 << s_{13}$, the Eq. (25) can be approximated as follow:

$$J_{\mathrm{CP}} \approx 0.2327 s_{13}\sin\delta. \quad (26)$$

If we insert the value of Jarlskog rephasing invariant of Eq. (22) into Eq. (26), then we have the Dirac phase δ as follow:

$$\delta \approx \arcsin\left(\frac{0.0172}{\sin\theta_{13}}\right). \quad (27)$$

By inserting the value of mixing angle θ_{13} as shown in Eq. (17), we have:

$$\delta \approx 11.2^o \quad (28)$$

4. Conclusions

We have studied systematically the effect of breaking on neutrino mass matrix that obey $\mu - \tau$ symmetry by introducing a complex parameter ix with the requirement that the trace of the broken $\mu - \tau$ symmetry is remain constant. By using the experimental data of mixing angles θ_{12} and θ_{23} as input, we can obtain the Jarlskog rephasing invariant $J_{\rm CP} \neq 0$ which indicate the existence of CP violation in neutrino sector and hence the Dirac phase δ which also depend on the mixing angle θ_{13} for neutrino mass in normal hierarchy for the case: $m_1 = 0$.

References

[1] Apollonio M *et al.* [Double Chooz Collab.] 1999 *Phys. Lett.* **B 466** 415
[2] Adamson P *et al.* [MINOS Collab.] 2011 *Phys. Rev. Lett.* **107** 181802
[3] Abe K *et al.* [T2K Collab.] 2011 *Phys. Rev. Lett.* **107** 041801
[4] An F P *et al.* 2012 *Phys. Rev. Lett.* **108** 171803
[5] Ahn J K *et al.* [RENO Collab.] 2012 *Phys. Rev. Lett.* **108** 191802
[6] Mohapatra R N 2004 *JHEP* **0410** 027
[7] Aizawa I and Yasue M 2006 *Phys. Rev.* **D 73** 015002
[8] Fuki F and Yasue M 2006 *Phys. Rev.* **D 73** 055014
[9] Mohapatra R N and Rodejohann W 2005 *Phys. Rev.* **D 72** 053001
[10] Baba T and Yasue M 2007 *Phys. Rev.* **D 75** 055001
[11] He H -J and Xu X -J 2012 Phys. Rev. **D 86**, 111301 (R)
[12] Damanik A 2013 Nonzero θ_{13}, CP violation, and $\mu - \tau$ symmetry *Preprint* arXiv:1305.6900 [hep-ph]
[13] Damanik A 2011 *Mod. Phys. Lett.* **A 26** 567
[14] Branco C G, Gonzalez Felipe R, Joaquim F R, Masina I, Rebelo M N, and Savoy C A 2003 *Phys. Rev.* **D 67** 073025
[15] Gonzales-Garcia M, Maltoni M and Salvado J 2010 *JHEP* **056** 1004
[16] Fogli G *et al.* 2010 *J. Phys. Con. Ser.* **203** 012103

Journal of Physics: Conference Series **539** (2014) 012013 doi:10.1088/1742-6596/539/1/012013

New development in radiative neutrino mass generation

Julio
Fisika LIPI, Kompleks Puspiptek, Tangerang 15310, Indonesia
E-mail: julio@lipi.go.id

Abstract. We present a simple and predictive model of radiative neutrino masses. It is a special case of the Zee model with a family-dependent Z_4 symmetry acting on the leptons. A variety of predictions follow: The hierarchy of neutrino masses must be inverted; the lightest neutrino mass is extremely small and calculable; one of the neutrino mixing angles is determined in terms of the other two; the phase parameters take CP-conserving values with $\delta_{CP} = \pi$; and the effective mass in neutrinoless double beta decay lies in a narrow range, $m_{\beta\beta}$ =(17.6–18.5) meV. The ratio of vacuum expectation values of the two Higgs doublets, $\tan\beta$, is determined to be either 1.9 or 0.19 from neutrino oscillation data. Flavor-conserving and flavor-changing couplings of the Higgs doublets are also determined from neutrino data. The non-standard neutral Higgs bosons, if they are moderately heavy, decay significantly into μ and τ with prescribed branching ratios. Observable rates for the decays $\mu \to e\gamma$ and $\tau \to 3\mu$ are predicted if these scalars have masses in the range of 150-500 GeV.

1. Introduction

The fact that neutrinos have masses and mix brings attentions to mechanisms that can explain it. The most popular way is seesaw mechanism in which right-handed neutrinos (which are singlets under Standard Model gauge groups) are integrated out resulting in dimension-five operator $LLHH/\Lambda$ with L and H being lepton and Higgs doublets, and Λ being right-handed neutrino mass. Once the Higgs acquires vacuum expectation value (VEV), this becomes neutrino mass term. The value of Λ, in accordance with neutrino oscillation data, is of order 10^{14} GeV. Such high scale is well beyond the reach of LHC, thus making it difficult to be tested experimentally.

There exists, however, class of models in which neutrino masses are generated at loop level. Because of loop and chirality suppressions, the new physics scale can be lower. Thus, they may be testable at the LHC as well as in lepton flavor violation processes. A well-known example is Zee model [1]. This model features two Higgs doublets $H_a(1,2,-1/2)$ $(a = 1,2)$ and scalar singlet $\eta^+(1,1,+1)$. The gauge-invariant Yukawa interactions in lepton sector can be written as

$$\mathcal{L}^{(\ell)}_{\rm Yuk} = \sum_{\substack{i,j=1,2,3,\\ a=1,2}} Y^a_{ij} L_i e^c_j H_a + \sum_{i,j=1,2,3} f_{ij} L_i L_j \eta^+ + h.c. \quad (1)$$

where i,j =1–3 indicate family indices. Fermi statistics implies that $f = -f^T$. In addition, there is also cubic term in scalar potential

$$V = \left\{\mu H_1 H_2 \eta^+ + h.c.\right\} + ... \quad (2)$$

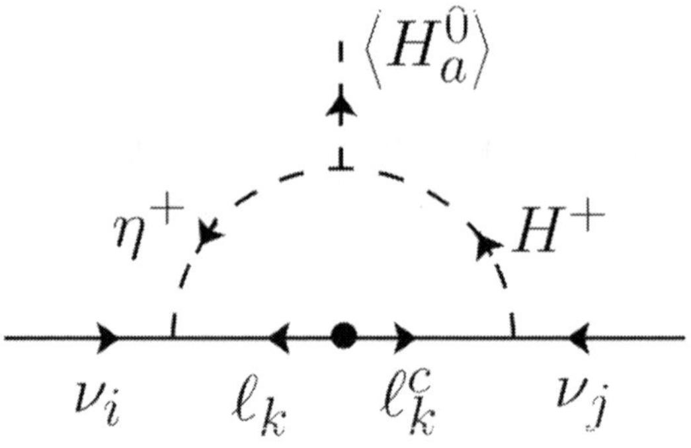

Figure 1. One–loop Zee model.

with "..." indicating potential terms not relevant for our discussion. The simultaneous presence of $f_{ij}L_iL_j\eta^+$ and $\mu H_1 H_2 \eta^+$ terms breaks lepton number by two units. The neutrino masses arise at one loop as shown in Fig. 1.

In its general form, both Higgs doublets are allowed to couple to all SM fermions. Consequently, the corresponding Yukawa matrices cannot be simultaneously diagonalized. This may lead to potentially dangerous tree-level flavor-changing neutral currents (FCNCs). On top of that, the structure of neutrino mass matrix is arbitrary in this case. Thus, it is hard to make a prediction for neutrino sector.

A more interesting approach was given by Wolfenstein in 1980. He imposed a Z_2 symmetry so that only one Higgs doublet could couple to fermions [2]. In contrast to the general Zee model, there is no FCNC induced at tree level here. One, therefore, can choose a basis where charged lepton Yukawa couplings are diagonal. The flavor structure of neutrino mass matrix suggests that all its diagonal elements must vanish identically. This implies that the neutrino mass hierarchy must be inverted and $\theta_{12} \simeq \pi/4$. However, this predicted value of θ_{12} is now ruled out by current solar and KamLAND data. These experiments prefer smaller value of θ_{12}. (See Refs. [3, 4] for detailed discussions.)

It is obvious that this Z_2 symmetry cannot describe a realistic neutrino mass and mixing. In this paper, we shall consider other discrete symmetry (i.e., Z_4 symmetry). But unlike Wolfenstein, our discrete symmetry is family dependent. As we shall show later, this approach leads to a predictive model of neutrino masses. All neutrino masses and mixings are described in terms of four real parameters. One neutrino mixing angle is given in terms of the other two. Neutrino mass hierarchy is predicted to be inverted. Tree-level FCNCs are induced, but their rates are consistent with current experimental bounds even for Higgs boson masses about few hundreds GeV.

2. Zee model with Z_4 discrete symmetry

We try to modify Zee model by introducing a family-dependent Z_4 symmetry as follows [5]:

$$\begin{aligned} &L_i : (-i,\, i,\, i); \quad\quad e_i^c : (-i,\, -i,\, -i); \\ &H_1 : +1; \quad H_2 : -1; \quad \eta^+ : -1\,. \end{aligned} \tag{3}$$

Here i=1–3 is the family index. Unlike the Wolfenstein realization, there will be tree-level FCNC induced here. But as we shall see later, they are within bounds allowed by experiments.

The leptonic Yukawa interactions, consistent with gauge and Z_4 symmetries can be written as:

$$\mathcal{L}_{\rm Yuk}^{(\ell)} = \sum_{\substack{i=2,3,\\ \alpha=1,2,3}} Y_{i\alpha} L_i e_\alpha^c H_1 + \sum_{\alpha=1,2,3} Y_\alpha L_1 e_\alpha^c H_2 + f_{23} L_2 L_3 \eta^+ + h.c. \tag{4}$$

Note that the cubic term necessary for breaking lepton number (Eq. (2)) is invariant under Z_4 symmetry.

In quark sector, it is necessary that all quarks couple universally to either H_1 or H_2:

$$\mathcal{L}_{\text{Yuk}}^{(q)} = \sum_{i,j=1-3} Y_{ij}^u Q_i u_j^c \tilde{H}_a + \sum_{i,j=1-3} Y_{ij}^d Q_i d_j^c H_a + h.c., \tag{5}$$

with $\tilde{H}_a = i\sigma_2 H_a^*$. With this form of Yukawa couplings it can be shown [5] that Z_4 charge assignment of Eq. (3) is anomaly-free [6,7].

It should be noted that one can always rotate the $L_{2,3}$ and $e_{1,2,3}^c$ fields such that $Y_{i\alpha}$ can be brought into diagonal form with Y_{22} and Y_{33} being nonzero. Such rotation will not induce any mixing in leptonic weak charged currents because both left-handed charged-leptons and neutrinos are rotated simultaneously. In this basis, the charged-lepton mass matrix can be written as

$$M_\ell = \begin{pmatrix} Y_1 v_2 & Y_2 v_2 & Y_3 v_2 \\ 0 & Y_{22} v_1 & 0 \\ 0 & 0 & Y_{33} v_1 \end{pmatrix}, \tag{6}$$

where v_1 and v_2 are vacuum expectation values (VEVs) of the two Higgses:

$$\langle H_1 \rangle = v_1, \quad \langle H_2 \rangle = v_2 = |v_2| e^{i\phi}. \tag{7}$$

Here v_1 has been made real by an $SU(2)_L \times U(1)_Y$ rotation.

We can rotate all phases in M_ℓ so that they become real. We then define

$$Y_1 v_2 = m_e^0 \sqrt{1+x^2+y^2}, \quad Y_2 v_2 = m_\mu^0 \frac{y\sqrt{1+x^2}}{\sqrt{1+x^2+y^2}}, \quad Y_3 v_2 = m_\tau^0 \frac{x}{\sqrt{1+x^2}},$$
$$Y_{22} v_1 = m_\mu^0 \frac{\sqrt{1+x^2}}{\sqrt{1+x^2+y^2}}, \quad Y_{33} v_1 = \frac{m_\tau^0}{\sqrt{1+x^2}}, \tag{8}$$

where $x, y, m_e^0, m_\mu^0, m_\tau^0$ are real parameters. Bringing the charged lepton fields into their mass eigenstates, the M_ℓ can be diagonalized by

$$M_\ell^{\text{diag.}} = U_L^T M_\ell U_R \simeq \text{diag.}(m_e^0, m_\mu^0, m_\tau^0), \tag{9}$$

with m_e^0, m_μ^0, m_τ^0 being approximate eigenvalues.

Among the two charged scalars H_1^+ and H_2^+, one combination forms a Goldstone boson $G^+ = (v_1 H_1^+ + v_2 H_2^+)/v$ (with $v \equiv \sqrt{v_1^2 + v_2^2}$). The orthogonal combination to it, $H^+ = (v_2^* H_1^+ - v_1 H_2^+)/v$, forms the physical charged scalar. In unitary gauge, one can write the couplings of H^+ with leptons (before rotations to lepton mass eigenbasis are performed) as

$$Y_{\text{Yuk}}^{(H^\pm)} = \frac{1}{v} \begin{pmatrix} -\frac{v_1}{v_2} & & \\ & \frac{v_2^*}{v_1} & \\ & & \frac{v_2^*}{v_1} \end{pmatrix} M_\ell\ . \tag{10}$$

In the mass eigenbasis of lepton fields, the matrix $Y_{\text{Yuk}}^{(H^\pm)}$ transforms into

$$\hat{Y} = \begin{bmatrix} \frac{m_e}{v}\frac{(x^2+y^2)\tan\beta-\cot\beta}{1+x^2+y^2} & -\frac{m_\mu}{v}\frac{y\,(\tan\beta+\cot\beta)}{\sqrt{1+x^2}\,(1+x^2+y^2)}, & -\frac{m_\tau}{v}\frac{x\,(\tan\beta+\cot\beta)}{\sqrt{1+x^2}\,\sqrt{1+x^2+y^2}} \\ -\frac{m_e}{v}\frac{y\,(\tan\beta+\cot\beta)}{\sqrt{1+x^2}\,(1+x^2+y^2)} & \frac{m_\mu}{v}\frac{\tan\beta\,\{(1+x^2)^2+x^2y^2\}-\cot\beta\, y^2}{(1+x^2)(1+x^2+y^2)} & -\frac{m_\tau}{v}\frac{xy\,(\tan\beta+\cot\beta)}{(1+x^2)\sqrt{1+x^2+y^2}} \\ -\frac{m_e}{v}\frac{x\,(\tan\beta+\cot\beta)}{\sqrt{1+x^2}\,\sqrt{1+x^2+y^2}} & -\frac{m_\mu}{v}\frac{xy\,(\tan\beta+\cot\beta)}{(1+x^2)\,\sqrt{1+x^2+y^2}} & \frac{m_\tau}{v}\frac{\tan\beta-x^2\cot\beta}{1+x^2} \end{bmatrix}, \tag{11}$$

where $\hat{Y} = U_L^T Y_{\text{Yuk}}^{(H^\pm)} U_R$ and $\tan\beta \equiv |v_2|/v_1$. Similarly, the flavor-antisymmetric matrix f is rotated into

$$\hat{f} = f_{23} \begin{pmatrix} 0 & \frac{x}{\sqrt{1+x^2}} & -\frac{y}{\sqrt{1+x^2}\sqrt{1+x^2+y^2}} \\ -\frac{x}{\sqrt{1+x^2}} & 0 & \frac{1}{\sqrt{1+x^2+y^2}} \\ \frac{y}{\sqrt{1+x^2}\sqrt{1+x^2+y^2}} & -\frac{1}{\sqrt{1+x^2+y^2}} & 0 \end{pmatrix}, \tag{12}$$

with $\hat{f} = U_L^T f U_L$.

The couplings of neutral scalar bosons to the leptons arise in analogous way. First, note that the pseudoscalar Higgs boson $A^0 = [v_2\,\text{Im}(H_1^0) - v_1\,\text{Im}(H_2^0)]/v$ would couple to the physical leptons as $\ell_i \hat{Y}_{ij} e_j^c A^0/\sqrt{2}$, where $\hat{Y}$ is the same Yukawa coupling matrix as in Eq. (11). For the other scalars, their couplings to lepton will involve another mixing angle α defined through $H^0 = \cos\alpha\,\text{Re}H_1^0 + \sin\alpha\,\text{Re}H_2^0$, $h^0 = -\sin\alpha\,\text{Re}H_1^0 + \cos\alpha\,\text{Re}H_2^0$. For the special choice $\alpha = \beta - \pi/2$, the neutral field h^0 will behave like the Standard Model Higgs field (i.e., its coupling to lepton fields are flavor diagonal), while H^0 couplings to lepton fields is nothing but $\hat{Y}$. This choice of α is realized in the decoupling limit, where the second Higgs doublet mass takes large values compared to v.

3. Phenomenology of neutrino mass

The neutrino masses are generated through one–loop diagram shown in Fig. 1. From that diagram, it is straightforward to determine the structure of neutrino mass matrix

$$M_\nu = \kappa(\hat{f} M_\ell^{\text{diag.}} \hat{Y}^T + \hat{Y} M_\ell^{\text{diag.}} \hat{f}). \tag{13}$$

Here, $M_\ell^{\text{diag.}}$ is the diagonal charged-lepton mass matrix, while $\hat{Y}$ and $\hat{f}$ are matrices given in Eqs. (11) and (12) respectively. The overall factor κ accounts for the loop integral,

$$\kappa = \frac{\sin 2\gamma}{16\pi^2} \log\left(\frac{M_1^2}{M_2^2}\right). \tag{14}$$

Note the main difference between our version and Wolfenstein model [2]. In the latter, $\hat{Y}$ in Eq. (13) is nothing but $M_\ell^{\text{diag.}}/v$, leading to vanishing of all M_ν diagonal elements.

Ignoring terms that are proportional to m_e^2 and m_μ^2 followed by diagonalizing the neutrino mass matrix, we obtain

$$U^T M_\nu U = M_\nu^{\text{diag}} = \text{diag.}\left\{\left(\frac{m_\tau^2 \kappa f_{23}}{v}\right)\frac{\sqrt{\tan^2\beta + x^2\cot^2\beta}}{1+x^2},\ -\left(\frac{m_\tau^2 \kappa f_{23}}{v}\right)\frac{\sqrt{\tan^2\beta + x^2\cot^2\beta}}{1+x^2},\ 0\right\} \tag{15}$$

where the matrix $U \equiv U_{\text{PMNS}}$, which is the Pontecorvo-Maki-Nakagawa-Sakata (PMNS) matrix, is found to be

$$U_{\text{PMNS}} = \begin{pmatrix} \frac{1}{\sqrt{2}}(C_\chi C_\psi + S_\psi) & \frac{1}{\sqrt{2}}(C_\chi C_\psi - S_\psi) & -S_\chi C_\psi \\ \frac{1}{\sqrt{2}}(C_\chi S_\psi - C_\psi) & \frac{1}{\sqrt{2}}(C_\chi S_\psi + C_\psi) & -S_\chi S_\psi \\ \frac{S_\chi}{\sqrt{2}} & \frac{S_\chi}{\sqrt{2}} & C_\chi \end{pmatrix}. \tag{16}$$

Here $S_\chi = \sin\chi$, $C_\chi = \cos\chi$ and $S_\psi = \sin\psi$, $C_\psi = \cos\psi$, with

$$S_\psi = \frac{y}{\sqrt{1+x^2+y^2}},\quad S_\chi = -\frac{(\tan\beta - x^2\cot\beta)}{\sqrt{1+x^2}\sqrt{\tan^2\beta + x^2\cot^2\beta}}. \tag{17}$$

The results obtained in Eqs. (15)-(16) suggest the following predictions. First, the hierarchy of neutrino masses must be inverted. This follows from the fact that (in the limit of $m_\mu = 0$) two neutrino mass eigenvalues are degenerate (up to signs) and another one is zero. Turning on the muon mass will remove the degeneracy and in the same time will induce the correct solar-mass splitting. Second, the Dirac phase δ_{CP} is predicted to be 0 or π. It comes from the reality property of neutrino mass matrix. Third, from the relation $|U_{\tau 1}| = |U_{\tau 2}|$ given in Eq. (16) it is found that θ_{23} can be expressed in terms of other mixing angles. Considering that θ_{23} must lie in the first quadrant and its value must be nearly 45 degrees, the only consistent solution is

$$t_{23} \quad = \quad s_{13}\frac{1+t_{12}}{1-t_{12}} \quad \text{and} \quad \delta_{CP} = \pi. \tag{18}$$

Here $t_{23} = \tan\theta_{23}$, etc. Thus the model predicts $\delta_{CP} = \pi$.

We plot the relation given in Eq. (18) in Fig. (2) in two planes: $\sin^2\theta_{23}$ versus $\sin^2\theta_{12}$ and $\sin^2\theta_{13}$ versus $\sin^2\theta_{23}$. As inputs we use $\sin^2\theta_{12}$ and $\sin^2\theta_{13}$ obtained from a global fit of neutrino data: $\sin^2\theta_{12} = 0.302 \pm 0.0125$ and $\sin^2\theta_{13} = 0.0227 \pm 0.0023$ both taken from [8]. In Fig. 2, we also show best fit to the mixing angles along with their error bars (shown in red). We can see that the prediction of the model is in very good agreement with the data although it is preferable for both $\sin^2\theta_{12}$ and $\sin^2\theta_{13}$ to be slightly above their central values by one sigma.

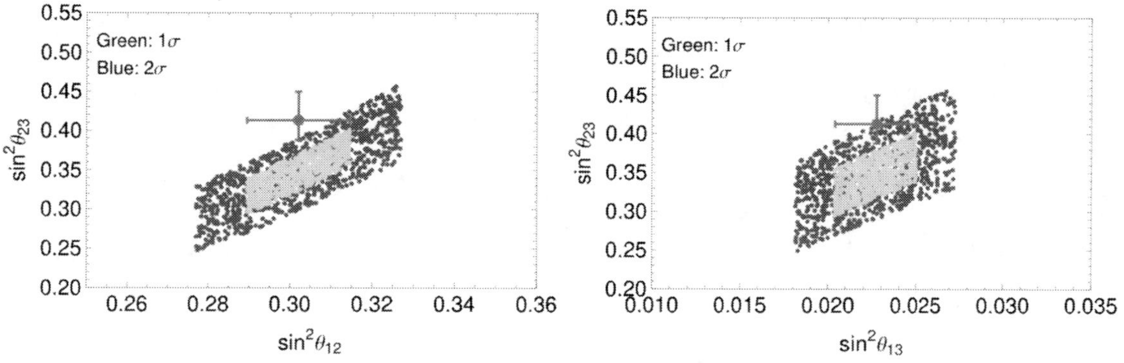

Figure 2. Predicted value of $\sin^2\theta_{23}$ as functions of $\sin^2\theta_{12}$ (left panel) and $\sin^2\theta_{13}$ (right panel).

The solar-mass splitting is generated by terms proportional to muon mass. Treating them as perturbations, we find

$$\Delta m^2_{\text{solar}} = 4\left(\frac{m_\mu^2}{m_\tau^2}\right)\frac{xy\tan\beta\sqrt{\tan^2\beta + x^2\cot^2\beta}}{(1+x^2+y^2)(\tan^4\beta + x^2)}\left(\frac{m_\tau^2\kappa f_{23}}{v}\right)^2, \tag{19}$$

and by using Eq. (15), we also get

$$\Delta m^2_{\text{atm}} \equiv m_3^2 - m_1^2 = -\frac{\tan^2\beta + x^2\cot^2\beta}{(1+x^2)^2}\left(\frac{m_\tau^2\kappa f_{23}}{v}\right)^2. \tag{20}$$

Now, the lightest neutrino mass m_3 is found to be $m_3 = \Delta m^2_{\text{solar}}/2|\Delta m^2_{\text{atm}}|^{1/2} \simeq 7.5\times 10^{-4}$ eV. The effective mass parameter $m_{\beta\beta}$ for neutrinoless double beta decay, therefore, can be determined from this relation. For this purpose, we use the atmospheric and solar mass splittings as input, obtained from the global fit: $\Delta m^2_{\text{atm}} = (2.47 \pm 0.07)\times 10^{-3}$ eV^2 and

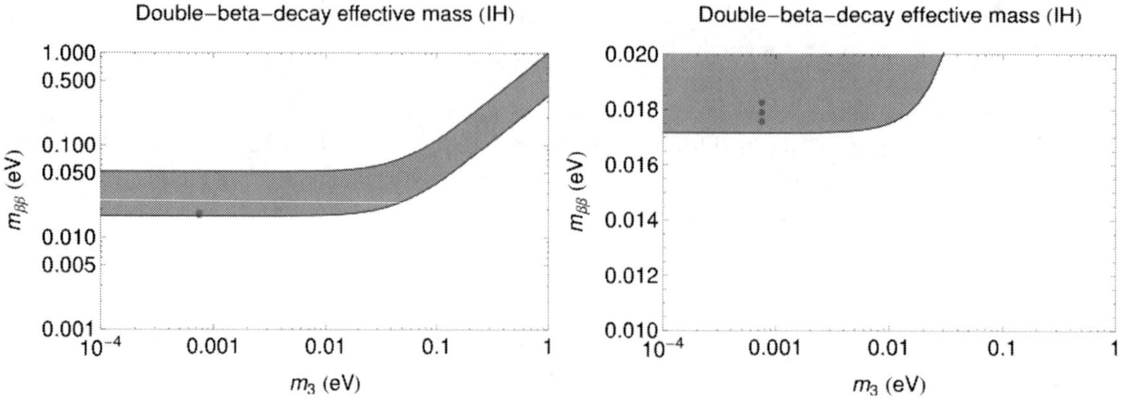

Figure 3. The range of double-beta-decay effective mass as function of the m_3 in inverted hierarchy (blue). The red points represent our prediction of $m_{\beta\beta}$ =17.6–18.5 meV. The zoomed region is shown in the right panel.

$\Delta m^2_{\rm solar} = (7.5 \pm 0.19) \times 10^{-5}$ eV^2. By varying the input parameters within their 1 sigma range, we found $m_{\beta\beta} \equiv |\sum_{i=1-3} U^2_{ei} m_i|$ =(17.6–18.5) meV. Here we also used the fact that the Majorana phases are zero, and that m_1 and m_2 have opposite CP eigenvalues. We also demand that the value of $\sin^2\theta_{23}$ resulting from the model prediction is within 1 sigma of the best fit value. As comparison, we plot the range of $m_{\beta\beta}$ (in IH case) as a function of the smallest neutrino mass in Fig. 3.

To determine the parameters of the model, we choose inputs values for $R = \Delta m^2_{\rm solar}/|\Delta m^2_{\rm atm}|$, $|U_{e2}|$ and $|U_{e3}|$. Using the relation for R from Eqs. (19)–(20), and the relations $|U_{e2}| = |\frac{1}{\sqrt{2}}(C_\chi C_\psi - S_\psi)|$ and $|U_{e3}| = |-S_\chi C_\psi|$, for a given input choice we solve for $(x,\, y,\, \tan\beta)$. The third mixing angle θ_{23} is determined through Eq. (18). We find two separate solutions for the parameters (up to signs):

$$\begin{aligned} &(i) \qquad (x,\, y,\, \tan\beta) = (0.038,\, 4.24,\, 0.19), \\ &(ii) \qquad (x,\, y,\, \tan\beta) = (4.6,\, 21.2,\, 1.9)\ . \end{aligned} \tag{21}$$

Solution (i) is obtained by using the analytic expressions with input values $(m^0_e,\, m^0_\mu,\, m^0_\tau) = (m_e, m_\mu, m_\tau)$, $|U_{e2}|^2 = 0.32$, $|U_{e3}|^2 = 0.0227$ and $R = 1/32.9$. We compared this result with the exact values determined numerically and found agreement within 1% for this case. For solution (ii), we improve the results obtained from analytic expressions iteratively by numerical methods. As input we choose $(m^0_e,\, m^0_\mu,\, m^0_\tau) = (1.001\, m_e,\, 1.041\, m_\mu,\, 0.9605\, m_\tau)$, $|U_{e2}|^2 = 0.324$, $|U_{e3}|^2 = 0.0202$ and $R = 1/32.3$. Both solutions fit neutrino oscillation data quite well. These values can now be used to compute the Higgs Yukawa coupling matrix $\hat{Y}$, which would determine the structure of flavor changing neutral currents.

For the two solutions we find this matrix to be

$$(i) \quad \hat{Y} = \begin{pmatrix} -2.95 \times 10^{-7} & -0.00074 & -0.00049 \\ -3.60 \times 10^{-6} & -0.003 & -0.002 \\ -1.40 \times 10^{-7} & -0.0001 & 0.0018 \end{pmatrix}, \tag{22}$$

$$(ii) \quad \hat{Y} = \begin{pmatrix} 5.57 \times 10^{-6} & -1.29 \times 10^{-5} & -0.0011 \\ -6.25 \times 10^{-8} & 0.0011 & -0.0046 \\ -3.22 \times 10^{-7} & -2.75 \times 10^{-4} & -0.0044 \end{pmatrix}. \tag{23}$$

With these coupling matrices, we can now determine FCNC rates, which we address in the next section.

The overall coefficient of neutrino mass matrix from Eq. (20) is found to be $\kappa f_{23} = \{9.8\times10^{-9},\ 2.1\times10^{-8}\}$ (for cases (i) and (ii)). The smallness of κf_{23} may be explained by choosing κ and/or f_{23} small. A small κ is realized if the cubic scalar coupling coefficient μ in Eq. (2) is small, or if the mass of one of the charged scalar η^+ or H^+ is large. As an illustration, choose $\mu = 1$ GeV, $f_{23} = 0.01$, $m_{\eta^\pm} = 1.5$ TeV and $m_{H^\pm} = 500$ GeV. This would yield $\kappa f_{23} = 2.1\times10^{-8}$, consistent with solution (ii). Clearly, other choices are also possible. This shows that the smallness of neutrino masses can be explained in the present framework without much tuning, even when the scale of new physics is near the TeV.

4. Lepton flavor violation and Higgs phenomenology

From the structure of $\hat{Y}$ given above, it is clear that there exist tree-level lepton flavor violation processes mediated by H^0 and A^0 scalars. In the limit $m_{H^0} = m_{A^0}$, we find the branching ratio of $\tau\to3\mu$ to be

$$(i)\quad \mathrm{BR}(\tau\to3\mu) = 1.6\times10^{-12}\,(150\ \mathrm{GeV}/m_{A^0})^4\,, \tag{24}$$

$$(ii)\quad \mathrm{BR}(\tau\to3\mu) = 1.2\times10^{-12}\,(150\ \mathrm{GeV}/m_{A^0})^4\,. \tag{25}$$

Other lepton flavor violation rates in this class are much suppressed.

On the other hand, radiative decays $\ell_i\to\ell_j+\gamma$ arise through one–loop diagrams mediated by the neutral Higgs bosons H^0 and A^0 as well as the charged Higgs bosons $H^\pm$ and $\eta^\pm$. Ignoring the small $H^+-\eta^+$ mixing (see the discussion in the last paragraph of Section 3) and taking $m_{H^0} = m_{A^0}$, we find their rates to be

$$\Gamma(\ell_i\to\ell_j\gamma) = \frac{\alpha_{em} m_{\ell_i}^5}{(96\pi^2)^2}\left[\frac{|\frac{1}{4}(\hat{Y}^T\hat{Y})_{ij}|^2}{m_{A^0}^4} + \frac{|\frac{1}{2}(\hat{Y}\hat{Y}^T)_{ij} - r^2(\hat{f}\hat{f}^T)_{ij}|^2}{m_{A^0}^4}\right], \tag{26}$$

where $r \equiv m_{A^0}^2/m_{\eta^+}^2$ and we have taken $m_{H^0} = m_{A^0}$. After substituting $(x, y, \tan\beta)$ values determined from neutrino oscillation data, it turns out that the last two terms interfere constructively. So, our estimates will be the lower limits of these processes. For $\mu\to e\gamma$, we find

$$(i)\quad \mathrm{BR}(\mu\to e\gamma) \geq 2.0\times10^{-15}\,(150\ \mathrm{GeV}/m_{A^0})^4\,, \tag{27}$$

$$(ii)\quad \mathrm{BR}(\mu\to e\gamma) \geq 5.1\times10^{-15}\,(150\ \mathrm{GeV}/m_{A^0})^4\,. \tag{28}$$

These are consistent with present limit and possibly within reach of MEG and other proposed experiments.

In quark sector, the Yukawa interactions when all quarks couple to H_1 are given by (in the limit $\alpha = \beta - \pi/2$)

$$\begin{aligned}\mathcal{L}_{\rm Yuk}^{(q)} &= -\sum_q \frac{m_q}{v\sqrt{2}}\bar{q}qh^0 + \tan\beta\sum_q\left[-\frac{m_q}{v\sqrt{2}}\bar{q}qH^0 + i\frac{m_q}{v\sqrt{2}}\bar{q}\gamma_5 qA^0\right]\\ &\quad -\tan\beta\frac{(V_{CKM})_{ij}}{v}\left[(m_{u_i}\bar{u}_iP_Ld_j - m_{d_j}\bar{u}_iP_Rd_j)H^+ + h.c.\right].\end{aligned} \tag{29}$$

When quarks couple to H_2 one just needs to replace $\tan\beta\to\cot\beta$, $A^0\to-A^0$, and $H^\pm\to-H^\pm$. Considering perturbativity, solution (i) can only be realized when all quarks couple to H_1 or else the charged Higgs coupling to top quark will be unperturbative. Conversely, solution (ii) suggests that all quarks must couple to H_2.

Now, we can determine partial decay rates for H^0, A^0, and $H^\pm$ into SM particles. Their values are found to be (for $m_{A^0} = m_{H^0} = 300$ GeV):

Solution (i): H_1 couples to quarks:

$$\mathrm{BR}(H^0 \to t^*\bar{t}) + \mathrm{BR}(H^0 \to t\bar{t}^*) = 0.15; \quad \mathrm{BR}(H^0 \to b\bar{b}) = 0.34; \quad \mathrm{BR}(H^0 \to \tau^+\tau^-) = 0.089;$$
$$\mathrm{BR}(H^0 \to \mu^+\mu^-) = 0.24; \quad \mathrm{BR}(H^0 \to \mu^+\tau^-) + \mathrm{BR}(H^0 \to \tau^+\mu^-) = 0.11; \tag{30}$$

Solution (ii): H_2 couples to quarks:

$$\mathrm{BR}(H^0 \to t^*\bar{t}) + \mathrm{BR}(H^0 \to t\bar{t}^*) = 0.23; \quad \mathrm{BR}(H^0 \to b\bar{b}) = 0.52; \quad \mathrm{BR}(H^0 \to \tau^+\tau^-) = 0.1;$$
$$\mathrm{BR}(H^0 \to \mu^+\mu^-) = 0.006; \quad \mathrm{BR}(H^0 \to \mu^+\tau^-) + \mathrm{BR}(H^0 \to \tau^+\mu^-) = 0.11; \tag{31}$$

Notice that there are no $H^0 \to W^+W^-, ZZ$ decays since the relevant couplings vanish in the limit of $\beta - \alpha = \pi/2$. In principle, H^0 could also decay into a pair of h^0. However, that coupling depends on a combination of quartic couplings which is unknown, and which may be very suppressed. We assume that this decay has a negligible rate. The charged-Higgs on the other hand, decays almost 100% of the time into $t\bar{b}$ in both solutions (i) and (ii).

In the present model, the cross section for H^0 (and A^0) production is suppressed by a factor of $(\tan\beta)^2$ in solution (i) and by a factor of $(\cot\beta)^2$ in solution (ii), compared to the cross section for a SM Higgs boson of identical mass. The cross sections for a SM Higgs boson of mass 300 GeV is 2.6 pb (3.6 pb) at $\sqrt{s} = 7$ TeV (8 TeV) [11]. Using this value, we find the model predictions for $pp \to (A^0, H^0) \to \mu^+\mu^-$ to be $\sigma B = (0.046\,pb, 0.008\,pb)$ for solutions (i) and (ii) at $\sqrt{s} = 7$ TeV and $\sigma B = (0.062\,pb, 0.012\,pb)$ for the two solutions at $\sqrt{s} = 8$ TeV. As a comparison, ATLAS collaboration has published their results on a search for heavy particles produced in pp collisions that decay into muon pairs. A limit of $\sigma B < 0.04\,pb$ has been obtained at 90% CL for the cross section times branching ratio of a sequential Z' decaying into muon pair, in an analysis using $5\,fb^{-1}$ of data collected at $\sqrt{s} = 7$ TeV [12]. While this limit will be modified somewhat for the case of a scalar resonance, we anticipate the limit to be not very different for the case of H^0 and A^0.

5. Conclusions

We have presented a special case of the general Zee model. We employed a family-dependent Z_4 symmetry that resulted in a total of four real parameters explaining the entire neutrino oscillation data. There is a variety of predictions in the neutrino sector. The CP violating parameter δ_{CP} is predicted to be π. Most interestingly, one of the neutrino oscillation angles is determined in terms of the other two angles. This nontrivial relation is found to be consistent with current data. The model employs two Higgs doublets and a charged singlet. A crucial prediction is the VEV ratio $\tan\beta$. We are able to determine its value from neutrino oscillation data. We found that $\tan\beta = 0.19$ or 1.9. The branching ratios of the neutral Higgs bosons of the model into fermions are then completely determined. We found that leptonic decays, involving the muon, can be significant, which can potentially raise the reach for such particles at the LHC. The charged and neutral Higgs bosons also mediate leptonic flavor violation, with $\mu \to e\gamma$ possibly within reach of proposed experiments. The decay $\tau \to 3\mu$, which arise at the tree-level is also significant. These lepton flavor violation with prescribed branching ratio could serve as another test for the model.

Acknowledgments

The author would like to thank K. S. Babu for valuable discussions. He is also grateful to the organizers of CTPNP 2014 for the hospitality during the conference. This work was supported in part by the Slovenian Research Agency and Riset Kompetitif LIPI 2014 titled "*Dinamika fluida pada energi tinggi*".

References

[1] Zee A 1980 Phys. Lett. B **93** 389 [Erratum-ibid 1980 B **95** 461]
[2] Wolfenstein L 1980 Nucl. Phys. B **175** 93
[3] Koide Y 2001 *Phys. Rev. D* **64** 077301
[4] He X G 2004 *Eur. Phys. J. C* **34** 371
[5] Babu K S and Julio J 2014 *Phys. Rev. D* **89** 053004
[6] Krauss L M and Wilczek F 1989 *Phys. Rev. Lett.* **62** 1221
[7] Ibanez L E and Ross G G 1991 *Phys. Lett. B* **260** 291
[8] Gonzalez-Garcia M C, Maltoni M, Salvado J, and Schwetz T 2012 *JHEP* **1212** 123
[9] Xing Z-z, Zhang H, and Zhou S 2008 *Phys. Rev. D* **77** 113016
[10] Babu K S 2009 arXiv:0910.2948 [hep-ph]
[11] LHC Cross Section Working Group, https://twiki.cern.ch/twiki/bin/view/LHCPhysics/CrossSections.
[12] Aad G *et al.* [ATLAS Collaboration] 2012, *JHEP* **1211** 138

CTPNP 2014: "From Universe to String's Scale" IOP Publishing
Journal of Physics: Conference Series **539** (2014) 012014 doi:10.1088/1742-6596/539/1/012014

Effective 8-Spinor Model for Leptons

Yunita Umniyati

Department of Mechatronics, Swiss German University, EduTown BSD City Tangerang 15339, Indonesia

E-mail: yunita.umniyati@sgu.ac.id

Abstract. The 8-spinor realization of the chiral Skyrme-Faddeev model to describe leptons is proposed. The lepton charge being identified with the Hopf index. Energy in the model is proved to be estimated from below by the Hopf index to the power of 3/4.

1. Introduction

The Skyrme's idea to describe baryons as topological solitons [1] proved to be fruitful in nuclear physics. In the Skyrme Model the topological charge $Q = \deg(S^3 \to S^3)$ is interpreted as the baryon number B. The similar idea to describe leptons as topological solitons was announced by Faddeev [2], in which the Hopf invariant Q_H is interpreted as the lepton number L. Proposed the unification of Skyrme and Faddeev models through effective 8-spinor field $\mathbf{\Psi}$ as suggested in [3].

If we take into account the existence of the special 8-spinors identity discovered by the Italian geometer Brioschi [4]:

$$2j_\mu j^\mu = s^2 + p^2 + \mathbf{v}^2 + \mathbf{a}^2 + \mathbf{\Delta}^2, \tag{1}$$

where the following quadratic spinor quantities are introduced:

$$s = \bar{\mathbf{\Psi}}\mathbf{\Psi},\ p = i\bar{\mathbf{\Psi}}\gamma^5\mathbf{\Psi},\ \mathbf{v} = \bar{\mathbf{\Psi}}\lambda\mathbf{\Psi},\ \mathbf{a} = i\bar{\mathbf{\Psi}}\lambda\gamma^5\mathbf{\Psi},\ j^\mu = \bar{\mathbf{\Psi}}\gamma^\mu\mathbf{\Psi},\ \tilde{j}^\mu = \bar{\mathbf{\Psi}}\gamma^\mu\gamma^5\mathbf{\Psi},$$

$$\Delta^2 = \left[(\varphi_1^+\varphi_1)(\varphi_2^+\varphi_2) - |\varphi_1^+\varphi_2|^2 + (\chi_1^+\chi_1)(\chi_2^+\chi_2) - |\chi_1^+\chi_2|^2\right] \geq 0.$$

where $\bar{\mathbf{\Psi}} = \Psi^+\gamma_0$ and λ standing for Pauli matrices in the flavor (isotopic) space. Here the diagonal (Weyl) representation for $\gamma_5 = \gamma_5^+$ is used and γ_μ, $\mu = 0, 1, 2, 3$, designate the unitary Dirac matrices acting on Minkowsky spinor indices.

8-spinors (as columns) are defined as

$$\Psi = \mathrm{col}(\psi_1, \psi_2),\ \psi_i = \mathrm{col}(\varphi_i, \chi_i),\ i = 1,\ 2,$$

with φ_i and χ_i being 2-spinors. The structure of the identity (1) leads to the natural conclusion that Higgs potential V in the effective spinor field model can be represented as the function of $j_\mu j^\mu$:

$$V = \frac{\sigma^2}{8}(j_\mu j^\mu - \varkappa_0^2)^2, \tag{2}$$

with σ and $\varkappa_0$ being some constant parameters. If the Hamiltonian of the model is bounded below by some positive function of the topological charge, it is possible to prove the existence of stationary soliton solutions endowed with topological charge and implementing energy minimum.

Note that depending on the choice of the boundary condition at $r \to \infty$ will be obtained different varieties contained in S^8 defined by (1).

2. Spinor realization of chiral models

To ensure the fulfillment of the condition $\bar{\mathbf{\Psi}}\lambda_3\mathbf{\Psi}_{\text{vac}} = \text{const}$, $Q_H = \pi_3(S^2)$, the easiest way by choosing a nonlinear spinor model with Lagrangian as follows:

$$\mathcal{L} = \mathcal{L}_1 + \mathcal{L}_2 + \mathcal{L}_3,$$

where:

$$\mathcal{L}_1 = \partial_\mu \bar{\Psi}\gamma^\nu j_\nu \partial^\mu \Psi, \tag{3a}$$

$$\mathcal{L}_2 = g_1[(\bar{\Psi}\partial_\mu\Psi)(\bar{\Psi}\partial_\nu\Psi)^* - (\bar{\Psi}\partial_\nu\Psi)(\bar{\Psi}\partial_\mu\Psi)^*]^2, \tag{3b}$$

$$\mathcal{L}_3 = -g_2[(\bar{\Psi}\lambda_3\psi) - \sigma_0)]^2. \tag{3c}$$

Energy–momentum tensor in this model is defined as:

$$T^\mu_\nu = \Pi^\mu_\Psi \partial_\nu \Psi + \partial_\nu \bar{\Psi}\Pi^\mu_{\bar{\Psi}} - \delta^\mu_\nu \mathcal{L}.$$

The energy density is not difficult to find with the basis of Euler's theorem for homogeneous functions, and energy is obtained by Noether's theorem.

Let the unit vector is determined as $\vec{n} = \frac{\vec{N}}{|\vec{N}|}$, where $\vec{N} = (\bar{\psi}\vec{\lambda}\psi)$ is Goldstone field. It can be shown that energy in this model is bounded below by Hopf index:

$$E > \alpha |Q_H|^{\frac{3}{4}},$$

as in Faddeev model.

From energy density, we find $[\partial_i|\vec{N}|^2\vec{n}^2 + |\vec{N}|^2(\partial_i\vec{n})^2] > |\vec{N}|^2(\partial_i\vec{n})^2$ or $\xi^\dagger\gamma^0 A\xi + \eta^\dagger\gamma^0 A\eta \geq \lambda_{\min}|\xi^\dagger\xi + \eta^\dagger\eta|$. Next, we consider the reference system in which $\mathbf{\Psi}_1 \neq 0$; $\mathbf{\Psi}_2 = 0$. With some algebras we find that:

$$(\partial_i N^a \partial_k N^b N^c \varepsilon^{abc})^2 = \{2(\bar{\Psi}_1\Psi_1)\partial_i(\bar{\Psi}_1\Psi_1)\partial_k(\bar{\Psi}_1\Psi_1)\}^2 \leq |\partial_i\Psi|^2|\Psi|^2|\partial_k\Psi|^2|\Psi|^2.$$

As the final result, it is shown that Hamiltonian in this model is bounded below by Hamiltonian Faddeev

$$E(\Psi, \partial_\mu\Psi) > \alpha(\nabla n^a)^2 + \beta f^2_{ik} > \text{const}|Q_H|^{\frac{3}{4}}.$$

3. The effective nonlinear model of 8-spinor field

In this paper we construct a Lagrangian density of our model, that is analog with Skyrme and Faddeev models:

$$\mathcal{L} = \frac{1}{2\lambda^2}\overline{D_\mu\Psi}\gamma^\nu j_\nu D^\mu\Psi + \frac{\varepsilon^2}{4}f_{\mu\nu}f^{\mu\nu} - V, \tag{4}$$

where λ and ε being constant parameters of the model. In (4) the extended covariant derivative is written as:

$$D_\mu = \partial_\mu - ie_0 A_\mu \Gamma_e. \tag{5}$$

Γ_e in (5) stands for the electric charge operator, with e_0 being constant interaction with the electromagnetic field. The standard form of Γ_e reads as follow:

$$\Gamma_e = \frac{1}{2}(\lambda_3 - I). \tag{6}$$

In second term of (4) $f_{\mu\nu}$ stands for the antisymmetric tensor of Faddeev-Skyrme type:

$$f_{\mu\nu} = \left(\bar{\psi}\gamma^{\alpha} D_{[\mu}\psi\right)\left(\overline{D_{\nu]}\psi}\gamma_{\alpha}\psi\right). \tag{7}$$

It should be stressed that the first term in (4) generalizes the $\sigma-$model term in Skyrme Model, and the second term gives the generalization of Skyrme (or Faddeev) term.

By taking $\varphi_i \to \chi_i$ and using approximate hydrogen substitution for separate angles, as well as an approximation of $|\Psi|^2 = 4[A\tanh(\alpha x)+B]$ in isotropic spherical coordinates $x = \log(r/r_0)$, we obtain functionals of energy $\mathcal{E}$ and topological charge Q_H in the implementation of the chiral spinor Skyrme- Faddeev model:

$$\begin{aligned}\mathcal{E} = \int_{-\infty}^{+\infty} dx \quad & \left\{ \frac{e^x}{\lambda^2} r_0 \left[\frac{1}{4}R'^2 + R^2(\Theta'^2 + \xi'^2 + \eta'^2 + 2\xi'\eta'\cos 2\Theta + 2\sin^2\Theta)\right] + \right.\\ + & \ 4\frac{\epsilon^2}{r_0} e^{-x} R'^2 R^2 \sin^2\Theta(\cos^2\Theta\cos^2 2\eta + \sin^2\Theta\cos^2 2(\xi-\eta)) + \\ + & \left. r_0^3\sigma^2 e^{3x}\left(R^2 - \frac{\varkappa_0^2}{4}\right)^2\right\}. \end{aligned} \tag{8}$$

To estimate the energy $\mathcal{E}$ we take into account $Q_H = 1$ and use trial function:

$$R = A\tanh(\alpha x) + B, \quad \Theta = 2\arctan(e^{-\alpha x}), \ \alpha \approx \sqrt{2}, \ \eta = 0, \tag{9}$$

where A and B are defined as:

$$A = \frac{1}{2}\left(\frac{\varkappa_0}{2} - R_0\right), \quad B = \frac{1}{2}\left(\frac{\varkappa_0}{2} + R_0\right),$$

After some algebras we obtain energy $\mathcal{E}$:

$$\mathcal{E} = \frac{r_0\alpha}{3\lambda^2}(5A^2 + 4B^2) + \frac{64A^2\alpha\epsilon^2}{15r_0}\left(\frac{A^2}{7} + B^2\right) + \frac{4r_0^3\sigma^2A^2}{3\alpha}\left[\frac{A^2}{5} + (A+2B)^2\right],$$

with α, r_0, and R_0 are parameters of minimization. By varying the energy functional on A provided $A + B = \varkappa_0/2$ and assuming that $\varkappa_0 \ll 1$ and $\sigma\lambda^2 \ll 1$, we find:

$$A \approx \frac{2}{9}\varkappa_0, \quad B \approx \frac{5}{18}\varkappa_0.$$

Further, by varying ε on r_0, we obtain estimation of the form:

$$r_0 \approx 0,3\varkappa_0\lambda\epsilon, \quad \mathcal{E} \approx \frac{\varkappa_0^3\epsilon}{6\lambda}.$$

4. Acknowledgments

I would like to thank Prof. Yuri Petrovich Rybakov for the continuous support of my research and writing this paper.

References

[1] Skyrme T H R 1962 *A Unified Field Theory of Mesons and Baryons* (Nucl. Phys. vol 31) pp 556-559
[2] Faddeev L D 1973 *Reports of Ac. of Sc. USSR* (Moscow) vol 210 pp 807-810
[3] Rybakov Y P 2011 *Phys. of Nuclei* vol 74 pp 1102-1105
[4] Cartan E 1938 *Leçons Sur la Theorié des Spineurs* (Paris) p 223

Journal of Physics: Conference Series **539** (2014) 012015 doi:10.1088/1742-6596/539/1/012015

On the minimizer of the Thomas-Fermi-Dirac-von Weizsäcker model

S. Wahyuni[1,2], W.S.B. Dwandaru[3], and M.F. Rosyid[1]

[1] Kelompok Penelitian Kosmologi, Astrofisika, dan Fisika Matematik (KAM), Jurusan Fisika FMIPA Universitas Gadjah Mada, Yogyakarta, Indonesia
[2] Jurusan Fisika FMIPA Universitas Negeri Semarang, Indonesia
[3] Jurusan Fisika FMIPA Universitas Negeri Yogyakarta, Indonesia

E-mail: yuniblr@yahoo.com, wipsarian@yahoo.com, farchani@ugm.ac.id

Abstract. The energy functional of Thomas-Fermi-Dirac-von Weizsäcker model with external potential is studied. The minimizer for the functional is investigated. Furthermore, the value and some properties of the minimizer are estimated from the functional without solving the associate Euler-Lagrange equation.

1. Introduction

The Thomas-Fermi (TF) theory was proposed by Thomas and Fermi independently in 1927. Dirac in 1930 introduced the exchange correction to the theory so that the corrected model is called the Thomas-Fermi-Dirac (TFD). Then the gradient correction of the kinetic energy was added by von Weizsäcker in 1935. Those theories were placed on a firm mathematical footing by Lieb (1981) [1]. Although the Thomas-Fermi-Dirac-von Weizsäcker (TFDW) theory has not been extensively studied as the other theories, there exist some authors who studied it relativistically, such as Engel and Dreizler (1987) who discuss a systematic construction of the relativistic version of the TFDW model on the basis of field theory (QED) [2]. Moreover, Müller, Engel and Dreizler (1989) also presented an extension of the relativistic TFDW model to systems in arbitrary time-independent external four-potentials [3].

The existence of a minimizer is an interesting theme to be investigated. The existence of the minimizer for TFW model was proved by Benguria, Brezis and Lieb in 1981 [4]. The nonexistence of a minimizer for TFDW model for electrons with no external potential has been investigated by Lu and Otto [5]. They realized that the TFDW functional energy without external potential has no minimizer when the number of electrons exceed a certain positive integer. They also gave an estimation that the value of the functional is in the interval $[0, (4/5)^3]$. This current work is a part of our attempt to understand the properties of the minimizer for TFDW model of electrons in the influence of an external potential, without "touching" the associate Euler-Lagrange equation. The main result of our work here is proven by following partly the proof of a proposition in the work of Lu and Otto [5].

The TFDW energy functional for electrons with external potential V is given as

$$E(\phi) := \int_{\mathbb{R}^3} \left[|\nabla\phi|^2 + F(\phi^2)\right] dx + D(\phi^2, \phi^2) + \int_{\mathbb{R}^3} V\phi^2 \, dx, \tag{1}$$

where $F(t) = t^{5/3} - t^{4/3}$ and $D(\cdot,\cdot)$ is the Coulomb interaction in $\mathbb{R}^3$, i.e.

$$D(f,g) = \int\int_{\mathbf{R}^3\times\mathbf{R}^3} \frac{f(x)g(y)}{|x-y|} dx\, dy.$$

The above functional can be rewritten as

$$E(\phi) := \int_{\mathbb{R}^3} \left[|\nabla\phi|^2 + \bar{F}(\phi^2)\right] dx + D(\phi^2,\phi^2), \tag{2}$$

where $\bar{F}(t) = t^{5/3} - t^{4/3} + Vt$.

The variational principle related to TFDW model is to find the minimizer of the TFDW functional energy, i.e. a function φ that minimizes $E(\phi)$ in the sense

$$E(\varphi) = \inf_{\phi\in X,\ \int\phi^2=m} E(\phi), \tag{3}$$

for suitable function space X.

2. The estimation of the minimizer

Suppose the above functional (1) has a minimizer and we denote it by φ.

Proposition *Let* $b(x) = \left(\frac{4+\sqrt{16-60V(x)}}{10}\right)^{3/2}$, *where* $V(x) \leq 4/15$ *for all* $x \in \mathbb{R}^3$. *Then we have* $\varphi(x) \in [0, b(x)]$ *for a.e.* $x \in \mathbb{R}^3$.

Proof. Since $E(|\varphi|) \leq E(\varphi)$, we may restrict to $\varphi \geq 0$. The $\text{argmin}_{t\geq 0}\bar{F}(t)$ is calculated as follows

$$\begin{aligned} \frac{\partial\bar{F}(t)}{\partial t} &= 0 \\ \frac{5}{3}t^{2/3} - \frac{4}{3}t^{1/3} + V &= 0 \end{aligned}$$

Let $t^{1/3} = \alpha$ then

$$\frac{5}{3}\alpha^2 - \frac{4}{3}\alpha + V = 0,$$

so we have the square root of the last equation as follows

$$\alpha_{1,2} = \frac{4\pm\sqrt{16-60V}}{10},$$

and

$$t_{1,2} = \left(\frac{4\pm\sqrt{16-60V}}{10}\right)^3. \tag{4}$$

Note that $V \leq 4/15$ ensures the value of critical points t to be a real number.

It is easy to show that the function $\bar{F}(t)$ assumes the minimum value at the critical point

$$t_1 = \left(\frac{4+\sqrt{16-60V}}{10}\right)^3. \tag{5}$$

Journal of Physics: Conference Series **539** (2014) 012015 doi:10.1088/1742-6596/539/1/012015

Take $\varphi_1 = \varphi \wedge \left(\frac{4+\sqrt{16-60V}}{10}\right)^{3/2}$ and let $m_1 = \int \varphi_1^2$. It is clear that $m_1 \le m$. Since φ minimizes equation (1), we have

$$E(\varphi) \le E(\varphi_1) + \inf_{\int \psi^2 = m - m_1} E(\psi). \tag{6}$$

Indeed, for any ψ with $\int \psi^2 = m - m_1$, let

$$\psi_L(x) = Z_L[\varphi_1(x) + \psi(x + Le_1)],$$

where $e_1 = (1, 0, 0) \in \mathbb{R}^3$ and Z_L is a normalization constant so that $\int \psi_L^2 = \int \varphi^2 = m$.

The value of the energy functional in ψ_L is given by

$$E(\psi_L) = \int_{\mathbb{R}^3} |\nabla \psi_L|^2 \, dx + \int_{\mathbb{R}^3} \bar{F}(\psi_L^2) \, dx + D(\psi_L^2, \psi_L^2). \tag{7}$$

The first term of (7) satisfies

$$\begin{aligned} \int_{\mathbb{R}^3} |\nabla \psi_L|^2 \, dx &= \int_{\mathbb{R}^3} |\nabla \left(Z_L[\varphi_1(x) + \psi(x + Le_1)] \right)|^2 \, dx \\ &\le Z_L^2 \left\{ \int_{\mathbb{R}^3} |\nabla \varphi_1(x)|^2 \, dx + \int_{\mathbb{R}^3} |\nabla \psi(x + Le_1)|^2 \, dx + \right. \\ &\quad \left. 2 \int_{\mathbb{R}^3} |\nabla \varphi_1(x)||\nabla \psi(x + Le_1)| \, dx \right\}. \end{aligned}$$

When $L \to \infty$, then $Z_L \to 1$, $\int_{\mathbb{R}^3} |\nabla \psi(x + Le_1)|^2 \, dx = \int_{\mathbb{R}^3} |\nabla \psi(x)|^2 \, dx$, and $\int_{\mathbb{R}^3} |\nabla \varphi_1(x)||\nabla \psi(x + Le_1)| \, dx \approx 0$ so that

$$\int_{\mathbb{R}^3} |\nabla \psi_L|^2 \, dx \le \int_{\mathbb{R}^3} |\nabla \varphi_1(x)|^2 \, dx + \int_{\mathbb{R}^3} |\nabla \psi(x)|^2 \, dx. \tag{8}$$

Since $\bar{F}(t) = F(t) + tV$ and $\int V \psi_L^2(x) \, dx = \int V \varphi_1^2(x) \, dx + \int V \psi_L^2(x) \, dx$ when $L \to \infty$, then the second term of equation (7) satisfies

$$\int \bar{F}(\psi_L^2) \, dx \le \int \bar{F}(\varphi_1^2(x)) \, dx + \int \bar{F}(\psi^2(x)) \, dx \tag{9}$$

The third term of (7) can be written as

$$\begin{aligned} D(\psi_L^2, \psi_L^2) &= \int\int \frac{\psi_L^2(x_1)\psi_L^2(x_2)}{|x_1 - x_2|^2} \, dx_1 dx_2 \\ &= Z_L^4 \int\int \frac{1}{|x_1 - x_2|^2} \left\{ \varphi_1^2(x_1)\varphi_1^2(x_2) + \varphi_1^2(x_1)\psi(x_2 + Le_1) + \right. \\ &\quad 2\varphi_1^2(x_1)\varphi_1(x_2)\psi(x_2 + Le_1) + \\ &\quad \psi^2(x_1 + Le_1)\varphi_1^2(x_2) + \psi^2(x_1 + Le_1)\psi^2(x_2 + Le_1) + \\ &\quad 2\psi^2(x_1 + Le_1)\varphi_1(x_2)\psi(x_2 + Le_1) + \\ &\quad [2\varphi_1(x_1)\psi(x_1 + Le_1)\varphi_1^2(x_2) + 2\varphi_1(x_1)\psi(x_1 + Le_1)\psi^2(x_2 + Le_1) + \\ &\quad \left. 4\varphi_1(x_1)\psi(x_1 + Le_1)\varphi_1(x_2)\psi(x_2 + Le_1)] \right\} \, dx_1 dx_2 \end{aligned}$$

Because of $\sigma(x)\psi(x + Le_1) = 0$ if $L \to \infty$ for every localized function σ, then

$$D(\psi_L^2, \psi_L^2) = D(\varphi_1^2, \varphi_1^2) + D(\psi^2, \psi^2) \tag{10}$$

Therefore, from equation (8), (9), and (10) we have

$$\limsup_{L\to\infty} E(\psi_L) = E(\varphi_1) + E(\psi) \tag{11}$$

We arrive at equation (6) as $E(\varphi) \leq E(\psi_L)$ for any L.

Consider the scaling $\psi_\eta = \eta^{-3/2}\psi(x/\eta)$, so $\int \psi_\eta^2 = \int \psi^2$. It is clear that $\psi_\eta \to 0$ as $\eta \to \infty$, as each term in equation (2) goes to zero. Therefore, we have $E(\varphi) \leq E(\varphi_1)$ as

$$\inf_{\int \psi^2 = m - m_1} E(\psi) \leq 0.$$

Suppose $|\{\varphi > \left(\frac{4+\sqrt{16-60V}}{10}\right)^{3/2}\}| > 0$, then as $\left(\frac{4+\sqrt{16-60V}}{10}\right)^3$ is the unique minimizer of $\bar{F}(t)$ for $t \geq 0$,

$$\int \bar{F}(\varphi_1^2) < \int \bar{F}(\varphi^2).$$

Note that $|\nabla\varphi_1| \leq |\nabla\varphi|$ and $D(\varphi_1^2, \varphi_1^2) < D(\varphi^2, \varphi^2)$, we arrive at a contradiction $E(\varphi_1) < E(\varphi)$. Therefore, $\varphi(x) \leq b(x)$ for $x \in \mathbb{R}^3$.

□

For small illustration, in the case of negative Coulomb potential $V(x) = -|x|^{-1}$ we have the following graph of $b(x)$.

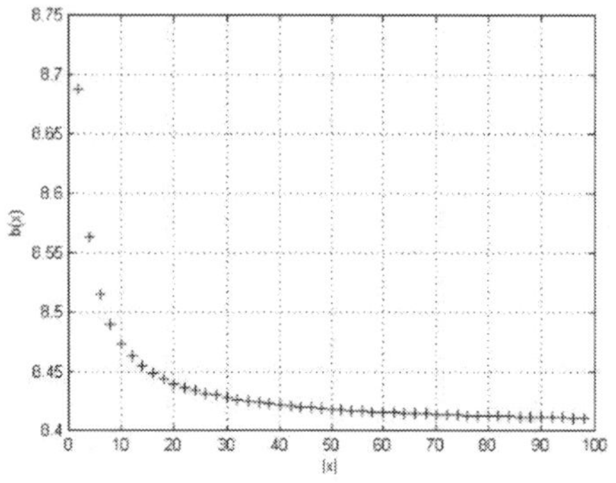

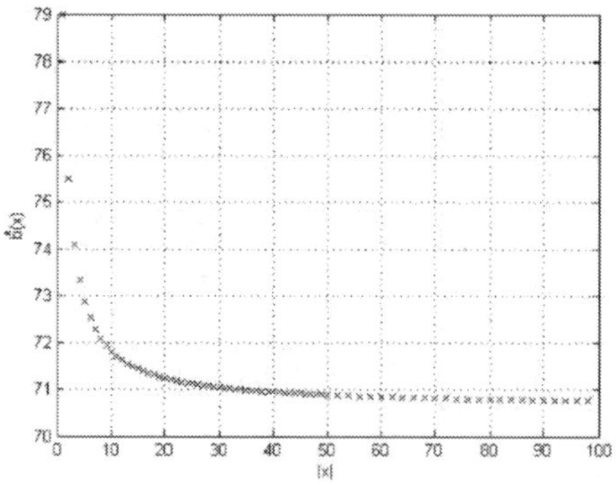

Figure 1. Left: The value of $b(x)$ with $V(x) = -|x|^{-1}$. Right: The value of $b^2(x)$ with $V(x) = -|x|^{-1}$.

3. Conclusion

The value of a minimizer is bounded from above by the "umbrella" function $b(x)$ which depends on the external potential $V(x)$.

4. References

[1] Lieb E H 1981 *Reviews of Modern Physics.* **53** 603
[2] Engel E and Drezler R M 1987 *Phys. Rev. A.* **35** 3607
[3] Müller H, Engel E and Drezler R M 1989 *Phys. Rev. A.* **40** 5542
[4] Benguria R, Brezis H and Lieb E H 1981 *Commun. Math. Phys..* **79** 167
[5] Lu J and Otto F 2013 Nonexistence of minimizer for Thomas-Fermi-Dirac-von Weizsäcker model *Comm. Pure Appl. Math.* doi: 10.1002/cpa.21477

Journal of Physics: Conference Series **539** (2014) 012016 doi:10.1088/1742-6596/539/1/012016

The equation of state of one-dimensional Gross-Pitaevskii equation

T. B. Prayitno

Physics Department, Faculty of Mathematics and Natural Science, Universitas Negeri Jakarta, Jl. Pemuda Rawamangun 13220, Indonesia

E-mail: trunk_002@unj.ac.id

Abstract. We have constructed the equation of state of one-dimensional Gross-Pitaevskii equation by benefiting from the formulation of a single macroscopic particle partition function and then modeling N undistinguishable macroscopic particles in canonical ensemble. This macroscopic particle, in which two or more atoms include, is literally a condensate that can be observed as an assembly in the system. In this case, we suppose all condensates are confined in the anisotropic parabolic trap and interaction between two condensates can be ignored by applying semi infinite cigar-shaped trap. It is also shown that the equation of state is indeed an ideal gas with the new thermodynamic interpretation of volume and pressure. Moreover, Even though our resulting partition function was portrayed as a sum of exponential functions, we prove that the series is indeed convergent.

1. Introduction

The phenomenon of Bose-Einstein condensation (BEC) has driven physicists to establish the new possibilities in order to utilize the condensed Bose gas. To produce the condensed Bose gas, one only has to cool down several atoms, such as alkali atoms [1, 2], below its specific critical temperature until the atomic waves overlap each other [3, 4]. It means that all atoms will occupy the lowest energy where some of them are condensed. Appropriate mathematical tools to characterize the mature of Bose gas in the very low temperature are in the most interest of this case. It has been noted that some authors have explored theoretically the properties of Bose gas, such as the soliton solution [5, 6, 7, 8, 9, 10], the propagation of atom laser wave [11, 12, 13, 14, 15, 16, 17], etc, by solving the Gross-Pitaevskii equation (GPE) which describes the dynamics of condensate or atom laser beam.

In thermodynamics, the property of gas can be understood by formulating the equation of state which provides the mathematical relation of two or more the thermodynamic quantities. The above equation is usually obtained by initially expressing the appropriate partition function which the assemblies can exchange both matter and energy. However, the problem to derive the partition function emerges in the nonlinear case since it is needed some assumptions and conditions. Nevertheless, in fact, in the case of BEC we have two options in order to deal with thermodynamic case by focusing on the particles or the condensates. In the previous papers, some authors proposed the grand canonical partition function to achieve some thermodynamic properties associated with N bosons confined by three-dimensional anisotropic harmonic trap [18, 19, 20, 21], for other similar discussions see also [22, 23]. They concerned with N weakly interacting particles by proposing a new variable thermodynamic quantity, namely harmonic volume, which replaces the role of real volume. In contrast to them, in order to achieve the thermodynamic quantities we consider the N noninteracting condensates, each of which is represented by the GPE, to formulate the canonical partition function by solving the GPE to obtain the energy levels. In our opinion, the most possible way to achieve the partition function simply is to consider the GPE as a macroscopic quantum oscillator since the trapping potential can be reduced to the harmonic potential Here, we adopt the idea by allowing the definition of harmonic volume since we also cannot find the available real volume. By pursuing our previous work, we offer the simplest way to attain the partition function by applying the canonical ensemble for N noninteracting condensates confined by the anisotropic harmonic trap. This model allows that a condensate may be treated as a macroscopic oscillator whose energy states are corrected on condition that the nonlinearity is considered as a small term. The derivation of the energy states, which were obtained by employing the time-independent perturbation theory, were proposed by Kivshar *et al.* [24], but we corrected the formulation in our previous paper where we only differ in the negative sign [25]. This textbook-like treatment generates a new perspective of the nonlinear modes presence like quantum mechanics version. In this paper, we present how to use the providing partition function to acquire the equation of state which relates to the pressure, volume and temperature. Besides, we also provide the other quantity, namely, the internal energy and heat capacity.

In the second section, we will be acquainting the canonical partition function of single condensate built by only considering the energy levels of macroscopic quantum oscillator, from which we extend for N noninteracting condensates. The convergence of the partition function is also explored by means of representing suitable function in the nonlinear term. In addition, we also introduce a new definition related to volume since the real volume is not provided in the partition function. We present in practice in section three how to derive the equation of state in the simplest way like in the textbook of statistical mechanics, as well as the internal energy and heat capacity. To convince us the validity of relation, we also prove that the thermodynamic relations are still hold. We close our work in section four by giving the conclusions.

2. Model of N noninteracting condensates partition function

We start by initially considering the one-dimensional GPE

$$i\frac{\psi}{\partial t}+\frac{\partial^2\psi}{\partial z^2}-z^2\psi+\sigma\left|\psi\right|^2\psi=0, \tag{1}$$

where the wave function ψ propagates in z direction and σ denotes the very small parameter in which the s wave scattering length has been absorbed, so σ can be negative or positive. It has been derived in our previous work by following up the former work of Kivshar *et al.* [24] the canonical partition function of the above equation of a single condensate [25]

$$\mathcal{Z}=\sum_{n=0}^{\infty}e^{\beta\left(\hbar\omega_z(n+1/2)-\sigma|B_n|^2V_{nnnn}\sqrt{m\omega_\perp\sqrt{\lambda}/\hbar}\right)}. \tag{2}$$

where we set $\lambda=\omega_z^2/\omega_\perp^2<<1$ in the cigar-shaped trap and define $\beta=-1/kT$. Here, we refer ω_z and $\omega_\perp$ to the longitudinal frequency and transverse frequency, respectively. The sum can be convergent or divergent based on the chosen function representing V_{nnnn}. The simplest method to choose the available function is linear least-square fit which provides linear combination of polynomials based on the list of values. If this method is applied, we note that the highest order of polynomials should be odd if considering the positive σ and even if σ is taken negative. For example, for listing ten data of V_{nnnn} versus n for the positive σ, by setting all entities to unity we are able to obtain the stable value of the sum, namely 9.75451 x 10^9, starting $n=42$. Later, by observing the formulation (1), it is clear that the partition function is constructed from the energy states of quantum oscillator containing a correction that comes from the nonlinearity

$$E_n = \left(n + \frac{1}{2}\right)\hbar\omega_z - \sigma \left|B_n\right|^2 V_{nnnn}, \tag{3}$$

where

$$V_{nmlk} = \int_{-\infty}^{\infty} \phi_n(z)\phi_m(z)\phi_l(z)\phi_k(z)\mathrm{d}z, \tag{4}$$

and ϕ_n is a set of dimensionless normalized eigen functions of harmonic oscillator

$$\phi_n(z) = (-1)^n \sqrt{\frac{1}{2^n n! \sqrt{\pi}}} e^{z^2/2} \frac{\mathrm{d}^n}{\mathrm{d}z^n}\left(e^{-z^2}\right). \tag{5}$$

This derivation was first presented by Kivshar *et al.* by treating the GPE in equation (1) as macroscopic quantum oscillator where the nonlinear term is considered as a very small term [24].

In order to obtain the equation of state in terms of pressure, temperature, and volume, we must fix the partition function in equation (2) to be an extensive entity. The problem is, there is no an extensive variable related to volume. Interestingly, this problem has been solved by Romero-Rochín by replacing the ordinary volume to harmonic volume which is the inverse cube of average geometric trapping frequency [18, 19, 20]

$$\mathcal{V} = \frac{1}{\omega^3}. \tag{6}$$

This introduced volume is motivated from the fact that an available volume in which gas fills at a given temperature, is proportional to $1/\omega^3$ [18]. Meanwhile, in the cigar-shaped trap the relation among the appropriate frequencies is given by [3]

$$\omega = \left(\omega_\perp^2 \omega_z\right)^{1/3}. \tag{7}$$

If we define the dimensionless parameter $\omega_\perp^2\omega_z/\omega^3$, then by observing the relation, we tranform the partition function with this parameter without loosing the meaning

$$\mathcal{Z} \rightarrow \left(\frac{\omega_\perp^2 \omega_z}{\omega^3}\right)\mathcal{Z} \tag{8}$$

or, in the complete form

$$\mathcal{Z} = \omega_\perp^2 \omega_z \mathcal{V} \sum_{n=0}^{\infty} e^{\beta\left(\hbar\omega_z(n+1/2) - \sigma|B_n|^2 V_{nnnn}\sqrt{m\omega_\perp \sqrt{\lambda}/\hbar}\right)}. \tag{9}$$

Since the dimensionless parameter can be set to unity, we claim then the thermodynamic potentials derived from the partition function does not lose the physical meaning. It is clear that our partition function now has become the extensive quantity since by transforming $\mathcal{V} \rightarrow \gamma\mathcal{V}$ with γ is an arbitrary positive constant, the partition function will also transform $\mathcal{Z} \rightarrow \gamma\mathcal{Z}$.

To build the noninteracting N condensates partition function, we have to choose whether the condensates form a set of distinguishable oscillators or indistinguishable oscillators. Nevertheless, we have examined that the treatment should be the second one since the set of N distinguishable oscillators leads to Gibbs paradox experienced by some thermodynamic quantities and it must be noted that the paradox is not available in the previous papers [18, 19, 20, 21, 22, 23] At this point, the total partition function of the set of N indistingushable condensates is given by

$$\mathcal{Z}_t = \frac{1}{N!}\left(\omega_\perp^2\omega_z\mathcal{V}\sum_{n=0}^{\infty} e^{\beta\left(\hbar\omega_z(n+1/2)-\sigma|B_n|^2V_{nnnn}\sqrt{m\omega_\perp\sqrt{\lambda}/\hbar}\right)}\right)^N. \tag{10}$$

In this case we assume that the number of condensates is so large, then we can use the Stirling's formula by agreeing the approximation

$$\frac{1}{N}\ln N! \cong \ln N - 1. \tag{11}$$

3. Equation of state

It is worth initially applying the partition function to derived one of the thermodynamic potentials, namely, the Helmholtz free energy

$$F = -kT\ln\mathcal{Z}_t. \tag{12}$$

Then, by substituting equation (10) into equation (12) and imposing the condition (11), we obtain

$$F = -NkT\ln\left(\frac{\mathcal{V}}{N}\omega_\perp^2\omega_z\sum_{n=0}^{\infty} e^{\beta\left(\hbar\omega_z(n+1/2)-\sigma|B_n|^2V_{nnnn}\sqrt{m\omega_\perp\sqrt{\lambda}/\hbar}\right)}\right) - NkT.$$

In statistical mechanics, the equation of state can be derived directly by using the equation of pressure. Since the pressure is indeed a pair conjugate of the volume, we call it as the harmonic pressure, $\mathcal{P}$, which is one of the intensive variables. In this case, by using the relation $\mathcal{P} = -(\partial F/\partial\mathcal{V})_{N,T}$, we find the equation state as an ideal gas form

$$\mathcal{P}\mathcal{V} = NkT. \tag{13}$$

It must be noted that the above equation of state can also be achieved with other thermodynamic relations, for example we are able to use the relation $\mathcal{P} = -(\partial F/\partial\mathcal{V})_{N,S}$, where E and S are internal energy and entropy, respectively.

However, even though the equation state has ideal gas formula, the internal energy and heat capacity are literally different. To show the differences, we use first the relation for the internal energy

$$\begin{aligned} E &= kT^2\frac{\partial\ln\mathcal{Z}_t}{\partial T} \\ &= N\frac{\sum_{n=0}^{\infty}\left[\left(n+\frac{1}{2}\right)\hbar\omega_z - \sigma\left|B_n\right|^2 V_{nnnn}\sqrt{\frac{m\omega_\perp\sqrt{\lambda}}{\hbar}}\right]\exp\left[-\left(n+\frac{1}{2}\right)\frac{\hbar\omega_z}{kT}+\sigma\frac{|B_n|^2}{kT}V_{nnnn}\sqrt{\frac{m\omega_\perp\sqrt{\lambda}}{\hbar}}\right]}{\sum_{s=0}^{\infty}\exp\left[-\left(s+\frac{1}{2}\right)\frac{\hbar\omega_z}{kT}+\sigma\frac{|B_s|^2}{kT}V_{ssss}\sqrt{\frac{m\omega_\perp\sqrt{\lambda}}{\hbar}}\right]} \end{aligned} \tag{14}$$

In order to plot the graphic E, it is convenient to replace V_{nnnn} in the numerator and denominator by a representing function. Realizing that V_{nnnn} cannot form recursion relation, we take ten data and employing linear least-square fit to represent V_{nnnn}. To end our discussion,

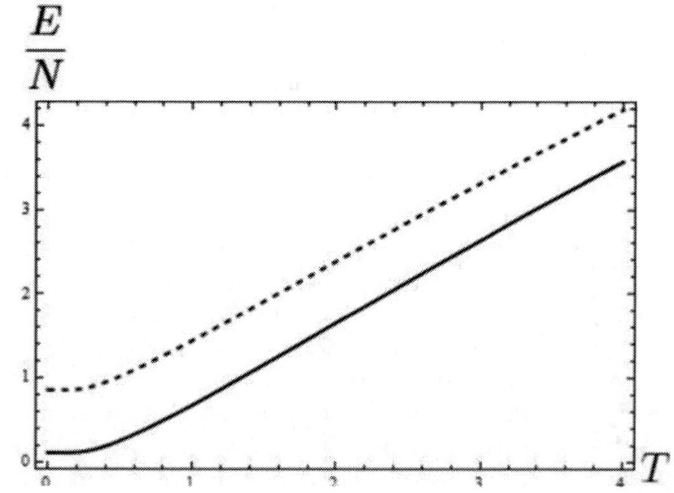

Figure 1. Internal energy per unit of number of condensates as a function of temperature for $\sigma = +$ shown by the solid line and $\sigma = -$ shown by the dotted line

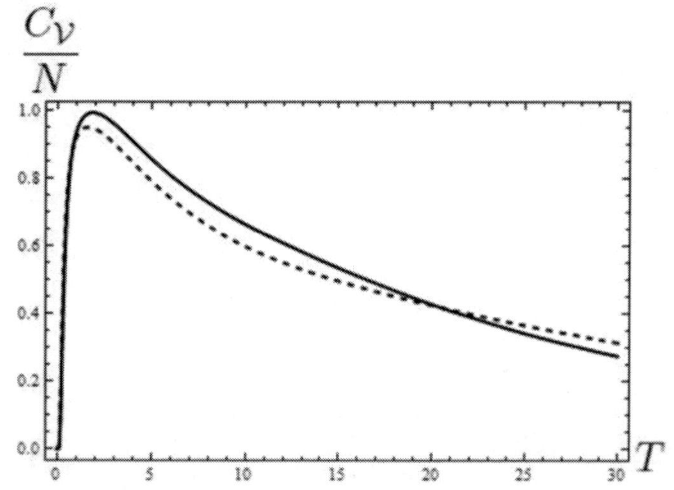

Figure 2. Heat capacity at constant volume vs temperature for $\sigma = +$ shown by the solid line and $\sigma = -$ shown by the dotted line

we attach another thermodynamic quantity, namely heat capacity at volume constant, which is derived from $C_\mathcal{V} = (\frac{\partial E}{\partial T})_{\mathcal{V},N}$. To produces all graphics, we replace V_{nnnn} with third order polynomial for positive σ and second order polynomial for negative σ, take $n = 42$, and set the other constants to unity. In figure 1, we show that raising temperature to the specific temperature in the interval of low temperature will not change the internal energy both positive and negative σ, and from that specific temperature the internal energy will linearly increase with the raising temperature. These results are not completely different with the ref. [19] since in the former paper the authors found that the internal energy is linear dependence with the temperature. Meanwhile, from another paper with the similar discussion [21] our results are fully different in calculating the heat capacity given in figure 2. In the previous paper [21],

Sandoval-Figueroa and Romero-Rochín found that at the specific temperature the heat capacity remains constant, while ours did not. We can see that at the very low temperature our heat capacity has a specific peak and monotonically decline from it.

4. Conclusions

As a conclusion, we confirm that the equation of state of N noninteracting undistinguishable condensates, which is confined by anisotropic parabolic trap, is derived by means of introducing $1/\omega^3$ as a harmonic volume to replace the ordinary volume in thermodynamics. This replacement is intended to discuss the thermodynamic entities through the partition function of a set of discrete energies of the one-dimensional GPE formulated by approaching time-independent perturbation theory. In addition, it also must be emphasized that we focus only on the dynamics of condensates, not the particles, by solving the GPE. As stated above, using canonical partition function to obtain the thermodynamic quantities will create the Gibbs paradox if we only consider the distinguishable condensate and the interaction of particles has been included explicitly in the GPE.

We also realize that this model causes the problem both in the internal energy and heat capacity, each of which is discontinuous at $T = 0$. Furthermore, in the case of heat capacity, we also cannot find the realistic entity since it must have a constant value in the higher temperature. This situation appears, in our opinion, since we neglect the interaction between condensates, minimally the attractive interaction.

4.1. Acknowledgments

It is a pleasure to acknowledge Dr. Anto Sulaksono for the encouragement and comments.

5. References

[1] Anderson M H, Ensher J R, Matthews M R, Wieman C E, Cornell E A 1995 *Science* **269** 198
[2] Davis K B, Mewes M -O, Andrews M R, van Druten N J, Durfee D S, Kurn D M, and Ketterle W 1995 *Phys. Rev. Lett.* **75** 3969
[3] Dalfovo F, Giorgini S, Pitaevskii L P, and Stringari S 1999 *Rev. Mod. Phys.* **71** 463
[4] Ketterle W 2002 *Rev. Mod. Phys.* **74** 1131
[5] Pérez-García V M, Michinel H, and Herrero H 1998 *Phys. Rev.* A **57** 3837
[6] Busch Th and Anglin J R 2001 *Phys. Rev. Lett.* **87** 010401
[7] Al Khawaja U, Stoof H T C, Hulet R G, Strecker K E, and Partridge G B 2002 *Phys. Rev. Lett.* **89** 200404
[8] Carr L D and Brand J 2004 *Phys. Rev. Lett.* **92** 040401 (*Preprint* cond-mat/0303257)
[9] Matuszewski M, Królikowski W, Trippenbach M, and Kivshar Y S 2006 *Phys. Rev.* A **73** 063621 (*Preprint* nlin/0609031)
[10] Atre R, Panigrahi P K, and Agarwal G.S 2006 *Phys. Rev.* E **73** 056611
[11] Schneider J and Schenzle A 1999 *Appl. Phys.* B **69** 353
[12] Bloch I, Hänsch T W, and Esslinger T 1999 *Phys. Rev. Lett.* **82** 3008
[13] Gerbier F, Bouyer P, and Aspect A 2001 *Phys. Rev. Lett.* **86** 4729
[14] Leggett A J 2001 *Rev. Mod. Phys.* **73** 307
[15] Busch Th, Köhl M, Esslinger T, and K. Mølmer 2002 *Phys. Rev.* A **65** 043615
[16] Cennini G, Ritt G, Geckeler C, and Weitz M 2005 *Phys. Rev. Lett.* **91** 240408
[17] Riou J -F, Guerin W, Le Coq Y, Fauquemberque M, Josse V, Bouyer P, and Aspect A 2006 *Phys. Rev. Lett.* **96** 070404
[18] Romero-Rochín V 2005 *Phys. Rev. Lett.* **94** 130601
[19] Romero-Rochín V and Bagnato V S 2005 *Brazilian. J. Phys.* **35** 607
[20] Romero-Rochín V 2005 *J. Phys. Chem.* **109** 21364
[21] Sandoval-Figueroa N and Romero-Rochín V 2008 *Phys. Rev.* E **78** 061129
[22] Sandoval-Figueroa N and Romero-Rochín V Thermodynamics and the virial expansion for trapped fluids in arbitrary external potentials *Preprint* cond-mat.stat-mech/0706.1322
[23] Romero-Rochín V The thermodynamic origin of the Contact and its relation to the gap in the BEC-BCS crossover *Preprint* cond-mat.quant-gas/1012.0236
[24] Kivshar Y S, Alexander T J, Turitsyn S K 2001 *Phys. Lett.* A **278** 225

[25] Prayitno T B and Latifa S 2012 *Makara J. Phys.* **16** 83

The application of path integral for log return probability calculation

DS Palupi[1], A Hermanto[1], E Tenderlilin[2] and MF Rosyid[1]

[1]Jurusan Fisika, Fakultas Matemtika dan Ilmu Pengetahuan Alam, Universitas Gadjah Mada,Indonesia
[2]Fakultas Ekonomika dan Bisnis, Universitas Gadjah Mada,Indonesia

E-mail: dwi_sp@ugm.ac.id

Abstract. Log return probability has been calculated using path integral method. The stock price is assumed obeying the stochastic differential equation of a geometric Brownian motion and the volatility is assumed following Ornstein Uhlenbeck process. The stochastic differential equation of stock price and volatility lead to Fokker-Plank equation. The Fokker-Plank equation is solved using path integral method. Distribution of log return can be used to take the valuation ln return stock.

1. Introduction

Study in financial market has been carried out by enorrmous authors in the framework of physics. One of those author is Scaden [1]. He modeled secondary financial market by making use of the machineries of quantum mechanics in which investors are repesented as vectors in a Hilbert space. All possible investors holding securities is taken as the basis of the Hilbert space of market states and the transactions are represented by operators. Bagarello [2] applied the above model introduced by Scaden in the case of finite dimensional investor Hilbert space.

In this work we calculate the log return probability using path integral method, in contrast to the work of Dragulescu and Yakovenko [3] in which they applied the characteristic method to calculate the same probability. The probability of log return plays some important roles in financial market analysis. At the first place the log return probability determines the valuation of the stock proper price. It also leads to the calculation of derivative price such as option and future.

We assumed that the stock price obeys the stochastic equation of a geometric Brownian motion, whereas the volatility is assumed following Ornstein-Uhlenbeck process. The assumption is also implemented in Heston's model [4]. The stochastic differential equation of stock price and volatility lead to a Fokker-Plank equation. The equation can be considered as Schroedinger-like equation and be solved using path integral method.

2. The Stochastic differential of stock price and volatility

Consider a stock whose price follows the stochastic differential equation of geometric Brownian motion as [3,4,5]

$$\mathrm{d}S(t) = \phi S(t)\mathrm{d}t + \sigma S(t)\mathrm{d}W_S, \tag{1}$$

where $S(t)$ is time function of stock price, ϕ drift parameter, and dW_S Wiener process standard. Drift parameter ϕ is assumed to be constant and risk free rate. Some author assumed that ϕ may depend on stock price, volatility as well as on time [6].

The volatility in equation (1) is assumed following Ornstein Uhlenbeck process [7]

$$\mathrm{d}\sigma(t) = -\eta\sigma(t)\mathrm{d}t + \mu \mathrm{d}W_v(t). \tag{2}$$

Using Ito's lemma, variance in equation (2) can be written as [4,6]

$$\mathrm{d}v(t) = -\gamma(v-\theta)\mathrm{d}t + \kappa\sqrt{v}\mathrm{d}W_v, \tag{3}$$

where $v = \sigma^2$ is variance. The factor $\mathrm{d}W_S$ and $\mathrm{d}W_v$ is Wiener process for stock price and Wiener process for variance respectively. Both of those processes have correlation given by

$$dW_v(t) = \rho dW_s(t) + \sqrt{1-\rho^2}Z(t)$$

with $\rho \in [-1,1]$.

There are some variables that can be investigated in financial [8]. One of the interesting variable to be analyzed is log return. In this work we would like to obtain the probability of log return. We use $z = l_r - \phi t$ as log return, with

$$l_r = \ln\frac{S(t)}{S(0)}.$$

Using Ito's calculus, equation (1) can be written as

$$\mathrm{d}z = -\frac{v}{2}\mathrm{d}t + \sqrt{v}\mathrm{d}W_S. \tag{4}$$

When two random variables X_i (i = 1,2) satisfy stochastic differential equations

$$\mathrm{d}X_i(t) = g_i(X_i(t))\mathrm{d}t + B_{ij}(X_i(t))\mathrm{d}W_j(t), \tag{5}$$

then we obtained the following Fokker-Plank equation [9,10]

$$\frac{\partial}{\partial t}P(x,t) = -\frac{\partial}{\partial x_i}\left[g_i(x,t)P(x,t)\right] + \frac{1}{2}\frac{\partial^2}{\partial x_i \partial x_j}\left[D_{ij}(x,t)P(x,t)\right], \tag{6}$$

where $D_{ij} = B_{ik}B_{jk}$ [9,10]. Equation (3) and (4) then lead to

$$\frac{\partial P}{\partial t} = -\gamma\frac{\partial}{\partial v}\left[(v-\theta)P\right] + \frac{1}{2}\frac{\partial}{\partial z}(vP) + \rho\kappa\frac{\partial^2}{\partial z \partial v}(vP) + \frac{1}{2}\frac{\partial^2}{\partial z^2}(vP) + \frac{\kappa}{2}\frac{\partial^2}{\partial z^2}(vP), \tag{7}$$

where $P = P(z,v\,|\,v_i)$ is the transition probability from initial state at $t = 0$ with log return $z = 0$ and variance v_i to the final state at time t with variance v and log return z. The initial condition for Fokker-Plank equation which governs time evolution of $P = P(z,v\,|\,v_i)$ is given by

$$P(t=0) = \delta(x)\delta(v-v_i) \quad . \tag{8}$$

The probability distribution of variance that satisfy equation (3) or Fokker-Plank equation for variance in equation (3) itself is given by

$$\frac{\partial \wp(t,v)}{\partial t} = \frac{\partial\left[\gamma(v-\theta)\wp(t,v)\right]}{\partial v} + \frac{\kappa^2}{2}\frac{\partial^2}{\partial v^2}\left[v\wp(t,v)\right]. \tag{9}$$

The solution of equation (9) in stationary state is

$$\wp_s(v) = \frac{\alpha^{\beta+1}}{\Gamma(\beta+1)}v^{\beta}\exp(-\alpha v), \tag{10}$$

where $\alpha = \frac{2\gamma}{\kappa^2}$ and $\beta = \alpha\theta - 1$. We can write equation (7) simpler via Fourier transformation

$$P(z,v \mid v_i) = \frac{1}{2\pi}\int_{-\infty}^{\infty} \mathrm{d}p_z e^{ip_z z}\tilde{P}(v \mid v_i) \tag{11}$$

as

$$\frac{\partial \tilde{P}}{\partial t} = \gamma\frac{\partial}{\partial v}\left[(v-\theta)\tilde{P}\right] - \left[\frac{p_z^2 - ip_z}{2}v - i\rho\kappa p_z\frac{\partial}{\partial v}v - \frac{\kappa^2}{2}\frac{\partial^2}{\partial v^2}v\right]\tilde{P}. \tag{12}$$

3.Path integral method for Fokker-Plank solution

Equation (12) can be considered as Schroedinger-like equation [5,11,12]

$$\frac{\partial \tilde{P}(v \mid v_i)}{\partial t} = -\hat{H}\tilde{P}(v \mid v_i) \tag{13}$$

which the Hamiltonian is given by

$$\hat{H} = \frac{\kappa^2}{2}\hat{p}_v^2\hat{v} - \mathrm{i}\gamma\hat{p}_v(\hat{v}-\theta) + \frac{p_z^2 - ip_z}{2}\hat{v} + \rho\kappa p_z\hat{p}_v\hat{v},$$

where $\hat{p}_v = -\mathrm{i}\frac{\partial}{\partial v}$ is the canonical operator conjugating to $\hat{v}$ which satisfy the commutation relation $[\hat{v},\hat{p}_v] = \mathrm{i}$. The solution of equation (13) are matrix elements of evolution operator

$$\tilde{P}(v,t \mid v_i,t_i) = \left\langle v_f \left| \exp(-\hat{H}t) \right| v_i \right\rangle, \tag{14}$$

if $\hat{H}$ is independent of time. Although the Hamiltonian $\hat{H}$ in equation (14.a) depends on time, this formula can be applied if the interval of time is very small so that $\hat{H}$ is nearly constant. So we can divide the time interval from t_i to t in equation (14.a) with M equal segments given by

$$\varepsilon = \frac{t - t_i}{M}$$

and then calculate $\tilde{P}(v,t \mid v_i,t_i)$ using path integral method. Now equation (14.a) can be expressed as

$$\tilde{P}(v,t \mid v_i,t_i) = \left\langle v_f \left| \exp(-\hat{H}t) \right| v_i \right\rangle = \int\prod_{j=1}^{M-1}\mathrm{d}v_j\prod_{j=1}^{M}\left\langle v_j \left| \exp\left(-\hat{H}(t_j - t_{j-1})\right) \right| v_{j-1} \right\rangle \quad . \tag{15}$$

Consider now the segment $(t_j - t_{j-1})$, inserting the completeness condition of basis $\{|p_{vj}\rangle\}$ into matrix elements $\left\langle v_j \left| \exp(-\hat{H}t) \right| v_{j-1} \right\rangle$ in this segment yields

$$\left\langle v_j \left| \exp\left(-\hat{H}(t_j - t_{j-1})\right) \right| v_{j-1} \right\rangle = \int \mathrm{d}p_{vj}\left\langle v_j \mid p_{vj}\right\rangle\left\langle p_{vj} \mid \exp\left(-\hat{H}(t_j - t_{j-1})\right) \mid v_{j-1}\right\rangle, \tag{16}$$

where

$$\left\langle v_j \mid p_{vj}\right\rangle = \frac{\exp(\mathrm{i}p_{vj}v_j)}{\sqrt{2\pi}}$$

and

$$\left\langle p_{vj} \mid v_{j-1}\right\rangle = \frac{\exp(-\mathrm{i}p_{vj}v_{j-1})}{\sqrt{2\pi}}.$$

Since ε is very small, we obtain

$$\left\langle p_{vj} \,|\, \exp\left(-\hat{H}\left(t_j - t_{j-1}\right)\right) | \, v_{j-1} \right\rangle \approx \exp\left(-\varepsilon H\left(p_{vj}, v_{j-1}\right)\right) \left\langle p_{vj} \,|\, v_{j-1} \right\rangle \tag{17}$$

and the probability $\tilde{P}(v,t\,|\,v_i,t_i)$ in equation (15) can be expressed as

$$\tilde{P}(v,t\,|\,v_i,t_i) = \int \prod_{j=1}^{M-1} \mathrm{d}v_j \int \prod_{j=1}^{M} \frac{\mathrm{d}p_{vj}}{2\pi}$$

$$\times \exp\left(\sum_{j=1}^{M}\left(\mathrm{i}p_{vj}v_j\right)\right) \exp\left(\sum_{j=1}^{M}\left(-\mathrm{i}p_{vj}v_{j-1}\right)\right) \exp\left(-\sum_{j=1}^{M-1} \varepsilon H\left(p_{vj}, v_{j-1}\right)\right), \tag{18}$$

where $H\left(p_{vj}, v_{j-1}\right) = \frac{\kappa^2}{2} p_{vj}^2 v_{j-1} - \mathrm{i}\gamma p_{vj}\left(v_{j-1} - \theta\right) + \frac{p_z^2 - \mathrm{i}p_z}{2} v_{j-1} + \rho\kappa p_z p_{vj} v_{j-1}$.

The path integral expression for equation (15) is obtained by taking limit $M \to \infty$ as [5,12]

$$\tilde{P}(v,t\,|\,v_i,t_i) = \int \mathrm{D}v\mathrm{D}p_v e^S \tag{19}$$

with the action is given by

$$S = \sum_{j=1}^{M}\left(\mathrm{i}p_{vj}\left(v_j - v_{j-1}\right) - \varepsilon\left(\frac{\kappa^2}{2} p_{vj}^2 v_{j-1} - \mathrm{i}\gamma p_{vj}\left(v_{j-1} - \theta\right) + \frac{p_z^2 - \mathrm{i}p_z}{2} v_{j-1} + \rho\kappa p_z p_{vj} v_{j-1}\right)\right). \tag{20}$$

Rearranging S as

$$S = \mathrm{i}p_{vM}v_M - \mathrm{i}p_{vi}v_i + \sum_{j=1}^{M-1} v_j\left[\mathrm{i}\left(p_{vj-1} - p_{vj}\right) - \varepsilon\frac{\kappa^2}{2} p_{vj}^2 + \mathrm{i}\varepsilon\gamma p_{vj} - \varepsilon\frac{p_z^2 - \mathrm{i}p_z}{2} - \epsilon\rho\kappa p_z p_{vj}\right]$$

$$-\sum_{j=1}^{M} \mathrm{i}\varepsilon\gamma p_{vj}\theta \tag{21}$$

and integrating equation (19) with respect to $\mathrm{d}v$ in interval j with S in equation (21) yields

$$\frac{\mathrm{d}p_v}{\mathrm{d}\tau} = i\frac{\kappa^2}{2} p_v^{\,2} + \Gamma p_v + i\frac{p_z^2 - ip_z}{2} \tag{22}$$

with $\Gamma = \gamma + i\rho\kappa p_z$ and the boundary condition $p_v = \tilde{p}_v$ at $\tau = t$.

The transition probability in equation (14) now remains

$$\tilde{P} = \int \frac{\mathrm{d}p_v}{2\pi} \exp\left[\mathrm{i}p_v v - \mathrm{i}p_{vi}v_i\right] \exp\left[-\mathrm{i}\int_0^t \gamma p_v \theta \mathrm{d}\tau\right]. \tag{23}$$

To obtain the solution of equation (23), we have to solve equation (22) firstly. Equation (22) can be expressed as

$$\frac{\mathrm{d}p_v}{\mathrm{d}\tau} = ap_v^{\,2} + bp_v + c, \tag{24}$$

with $a = \frac{\mathrm{i}\kappa^2}{2}, b = \Gamma, c = \frac{\mathrm{i}}{2}\left(p_z^{\,2} - ip_z\right)$ [13].By substituting $p_v = x + r$ into (24), where r is root of $ap_v^{\,2} + bp_v + c = 0$, i.e. $r = \frac{-\Gamma \pm \Omega}{\mathrm{i}\kappa^2}$, and $\Omega = \sqrt{\Gamma^2 + \kappa^2\left(p_z^{\,2} - \mathrm{i}p_z\right)}$ we obtain

$$\frac{\mathrm{d}x}{\mathrm{d}\tau} - \left(b + 2ar\right)x = ax^2. \tag{25}$$

Equation (25) is Bernoulli equation

$$\frac{\mathrm{d}y}{\mathrm{d}u} + P(u)y = g(u)y^n,$$

with $P=-(b+2ar)$, $g(u)=a$, and $n=2$. There are two kinds of r so we have two solutions of equation (25). Inserting the solutions into $p_v=x+r$ yields

$$p_v=-\frac{i2\Omega}{\kappa^2}\frac{1}{\chi_+\exp(\Omega(t-\tau))-1}+\mathrm{i}\frac{\Gamma-\Omega}{\kappa^2} \quad (26.a)$$

and

$$p_v=-\frac{i2\Omega}{\kappa^2}\frac{1}{1-\chi_-\exp(\Omega(\tau-t))}+\mathrm{i}\frac{\Gamma+\Omega}{\kappa^2} \quad (26.b)$$

where χ_+ and χ_- are

$$\chi_+=\left(1-\frac{2\Omega\mathrm{i}}{((\tilde{p}_v)\kappa^2-i(\Gamma-\Omega))}\right)$$

and

$$\chi_-=\left(\frac{\mathrm{i}2\Omega}{(\kappa^2\tilde{p}-i(\Gamma+\Omega))}+1\right)$$

respectively.

Inserting equation (26.a) and (26.b) into (23) then integrating these equations with respect to τ yields

$$\tilde{P}(v\,|\,v_i)=\int\frac{\mathrm{d}p_v}{2\pi}\exp[ip_v v-ip_{vi}v_i]\exp\left[-\frac{2\gamma\theta}{\kappa^2}\ln\frac{\chi_+-e^{-\Omega t}}{\chi_+-1}+\frac{\gamma\theta(\Gamma-\Omega)}{\kappa^2}t\right] \quad (27.a)$$

and

$$\tilde{P}(v\,|\,v_i)=\int\frac{\mathrm{d}p_v}{2\pi}\exp[ip_v v-ip_{vi}v_i]\exp\left[-\frac{2\gamma\theta}{\kappa^2}\ln\frac{(e^{\Omega t}-\xi_-)}{(1-\xi_-)}+\frac{\gamma\theta(\Gamma+\Omega)t}{\kappa^2}\right] \quad (27.b)$$

The probability $P(z,v\,|\,v_i)$ obtained by inserting equation (27.a) and (27.b) into Fourier transform (11) is given by

$$P(z,v\,|\,v_i)=\int\int_{-\infty}^{\infty}\frac{\mathrm{d}p_z}{2\pi}\frac{\mathrm{d}p_v}{2\pi}e^{\mathrm{i}p_z z}\exp[\mathrm{i}p_v v-\mathrm{i}p_{vi}v_i]\exp\left[-\frac{2\gamma\theta}{\kappa^2}\ln\frac{\chi_+-e^{-\Omega t}}{\chi_+-1}+\frac{\gamma\theta(\Gamma-\Omega)}{\kappa^2}t\right] \quad (28)$$

and

$$P(z,v\,|\,v_i)=\int\int_{-\infty}^{\infty}\frac{\mathrm{d}p_z}{2\pi}\frac{\mathrm{d}p_v}{2\pi}e^{\mathrm{i}p_z z}\exp[\mathrm{i}p_v v-\mathrm{i}p_{vi}v_i]\exp\left[-\frac{2\gamma\theta}{\kappa^2}\ln\frac{(e^{\Omega t}-\chi_-)}{(1-\chi_-)}+\frac{\gamma\theta(\Gamma+\Omega)t}{\kappa^2}\right] \quad (29)$$

Variance and volatility can not be obtained from financial data directly, and we interest with the log return whatever the volatility, so we concentrate to $P(z)$. Integrating equation (28) and (29) with respect to v and p_v respectively and applying $p_v=0$ at $\tau=t$ and $p_v(\tau=t)=\tilde{p}_v$, yields the same expression

$$P(z\,|\,v_i)=\int_{-\infty}^{\infty}\frac{dp_z e^{\mathrm{i}p_z z}}{(2\pi)}\exp\left[\frac{-(p_x^2-\mathrm{i}p_x)v_i}{\Gamma+\Omega\coth\left(\frac{\Omega t}{2}\right)}\right]\exp\left[-\frac{2\gamma\theta}{\kappa^2}\ln\frac{\left(\Gamma\sinh\left(\frac{\Omega t}{2}\right)+\Omega\cosh\left(\frac{\Omega t}{2}\right)\right)}{\Omega}+\frac{\gamma\theta\Gamma t}{\kappa^2}\right] \quad (30)$$

The probability $P(z)$ can not be obtain unless we have the distribution of v_i. When the distribution of v_i at $t = 0$ is stationer we can use equation (10) for v_i as

$$\wp_s(v_i) = \frac{\alpha^{\beta+1}}{\Gamma(\beta+1)} v_i^{\beta} \exp(-\alpha v_i),$$

and then $P(z)$ will be obtained from

$$P(z) = \int_0^\infty \mathrm{d}v_i \wp_s(v_i) P(z \mid v_i) \tag{31}$$

by integrating with respect to v_i as

$$P(z) = \frac{1}{(2\pi)} \int_{-\infty}^{\infty} dp_z e^{ip_z z} \exp\left(\frac{\gamma\theta\Gamma t}{\kappa^2} - \frac{2\gamma\theta}{\kappa^2} \ln\left(\frac{(\Omega^2 - \Gamma^2) + \Gamma 2\gamma}{\Omega 2\gamma} \sinh\frac{\Omega t}{2} + \cosh\left(\frac{\Omega t}{2}\right)\right)\right). \tag{32}$$

Equation (32) can be integrated easily by numeric.

4. Conclusion.

The probability of log return $P(z)$ can be calculated using path integral in which the initial volatility is set in stationary state. $P(z)$ is obtained from $P(z, v \mid v_i)$, and $P(z, v \mid v_i)$ is solution of Fokker Plank two dimensional that is built by stochastic differential equation for log return and volatility. $\tilde{P}(v \mid v_i)$ that is fourier transformation for $P(z, v \mid v_i)$ is calculated using path integral.

References

[1] Schaden M 2002 *Preprint* arXiv:physics/0203006
[2] Bagarelo F 2006 *Phys.A.* **39** 0823
[3] Dragulescu A A and Yakovenko V M 2002 *Quant. Finance* **2** 443
[4] Heston S L 1993 *Rev. Financial studies* **6** 327
[5] Hull J and White A 1987 *J. Finance* **42** 281
[6] Baaqui *Quantum Finance , Path Integral and Hamiltonian for Options and Interest Rates* B E 2004 (New York:Cambridge University Press)
[7] Stein E M and Stein J C 1991 *Rev. Financial studies* **4** 727
[8] Mantegna R N and Stanley H E 2000 *An Introduction to Econopysics, Corelation and Compexity in Finance* (Cambridge: Cambridge University Press)
[9] Breuer H P and Petruccione F 2003 *The Theory of Open Quantum system* (Oxford: University Press)
[10] Paul W and Baschanagel J 2000 *Stochastic Processes from Physics to Finance* (Berlin: Spinger-Verlag)
[11] Sakurai J J 1994 *Modern Quantum Mechanics* ed *Tuan S F* (New York:Addison-Wesley)
[12] Feynman R P and Hibbs A R 2010 *Quantum Mechanics and Path Integral* emended edition ed Styer DF (New York: Dover)
[13] Campbell S L 1990 *An Introduction to Differential Equation and Their Applications* (California:Wardworth Publishing Company)

Journal of Physics: Conference Series **539** (2014) 012018 doi:10.1088/1742-6596/539/1/012018

Approximate Solution of Schrodinger Equation in D-Dimensions for Scarf Hyperbolic plus Non-Central Poschl-Teller Potential Using Nikiforov-Uvarov Method

U.A. Deta[a,1], A. Suparmi[b], and C. Cari[b]

[a]Department of Physics, Faculty of Mathematics and Natural Science
Surabaya State University, Indonesia
Jl. Ketintang, Surabaya 60231, Indonesia

[b]Physics Department of Graduate Program
Sebelas Maret University, Indonesia
Jl. Ir. Sutami 36A Kentingan, Surakarta 57126, Indonesia

[1]E-mail: utamaalan@yahoo.co.id

Abstract. The approximate analytical solution of Schrodinger equation in D-Dimensions for Scarf hyperbolic plus non-central Pocshl-Teller potential were investigated using Nikiforov-Uvarov method. The approximate bound state energy are given in the close form and the corresponding approximate wave function for arbitary l-state in D-dimensions are formulated in the form of generalized Jacobi Polynomials. Special case is given for the ground state in 3 dimensions. The existence of arbitrary dimensions increase bound state energy system. In the other hand, the existence of arbitrary dimensions decreases the amplitude of wave function. The effect of Scarf Hyperbolic potential increases the bound state energy of system. The effect of non central Poschl-Teller potential decreases the bound state energy of system.

1. Introduction

The analytical solutions of Schrodinger equations for some physical potentials are very essential since they provide the important information of the quantum system. Recently, considerable efforts have been paid to obtain the exact solution of the shape invariant potentials. These potentials include Coulomb, Morse, Poschl-Teller, Hulten, inverted generalized hyperbolic potentials. The bound state energy spectra of these potentials have been investigated by various techniques such as Coulomb potential using Laplace transformation [1], Morse potential using series expansion method [2], Modified Poschl-Teller, Hulten, and Scarf hyperbolic and Scarf hyperbolic potential using NU method [3-6]. However, some shape invariant potentials in D-dimensions resolved only in radial part [1-6]. The angular part of wave function which have some potentials still unresolved.

The exact solution of Schrodinger equation is obtained if the angular momentum $l = 0$ in 1-dimension. Nevertheless, for $l \neq 0$ and $D > 1$, the Schrodinger equation can only be solved approximately for

[1] Correspondence Author.

Journal of Physics: Conference Series **539** (2014) 012018 doi:10.1088/1742-6596/539/1/012018

different suitable approximation scheme. One of the suitable approximation scheme is conventionally proposed by Greene and Aldrich [7,8].

In this paper we will attempt to solve the Schrodinger equation in D-Dimensions using Nikivorof Uvarov method. The NU method which was developed by Nikiforov-Uvarov [9]. This method based on solving the second order linear differential equations by reducing it to a generalized equation of hypergeometric type by a suitable change of variable.

Recently, the radial part of Schrodinger equation in D-Dimensions have been solved with various method. The combination between radial plus angular potential still haven't been studied yet. One of the combination shape invariant potential in D-Dimensions that remain unresolved is Scarf hyperbolic plus non-central Poschl-Teller potential. The Scarf potential describe particles which periodically arranged such as a crystal [5,6,10,11]. The application of solution in this potential are crystal model in solid state physics [5,6]. The Poschl-Teller potential describe the particle in Poschl-Teller Oscilattor. Moreover, the modified Poschl-Teller potential can been used to derive the well-known SO(2) spectrum generating algebra foran infinite square well problem [12].

The solution Scarf hyperbolic and Poschl-Teller potential in D-Dimensions still remain in radial part [3,5,6]. In the combined potential in D-Dimensions with the centrifugal term, this potential is separable potential, therefore, it can be solved using variable separation method. The radial part of Schrodinger equation have one solution. Nevertheless, the angular part of Schrodinger equation have D-1 solutions.

This paper is organized as follows. In section 2, we review the Nikiforov-Uvarov (NU) method briefly. The Scarf Hyperbolic plus Non-Central Pocshl-Teller Potential and their solution described in the section 3. In the section 4, special case given for the ground and first excited state in 3 dimensions. A brief results and conclusion in section 5.

2. Review of Nikiforov-Uvarov Method

The D-dimensional Schrodinger equation of any shape invariant potential can be reduced into hypergeometric or confluent hypergeometric type differential equation by suitable variable transformation [13-16]. The hypergeometric type differential equation, which is solved using Nikiforov-Uvarov method, is presented as:

$$\frac{\partial^2\Psi(s)}{\partial s^2}+\frac{\bar{\tau}(s)}{\sigma(s)}\frac{\partial\Psi(s)}{\partial s}+\frac{\bar{\sigma}(s)}{\sigma^2}\Psi(s)=0 \tag{1}$$

where $\sigma(s)$ and $\bar{\sigma}(s)$ are polynomials at most in the second order, and $\bar{\tau}(s)$ is first order polynomial. Equation (1) can be solved using separation of variable method which is expressed as:

$$\Psi=\phi(s)y(s) \tag{2}$$

By inserting equation (2) into equation (1) we get hypergeometric type equation, that is:

$$\sigma\frac{\partial^2 y}{\partial s^2}+\tau\frac{\partial y}{\partial s}+\lambda y=0 \tag{3}$$

$\phi\ (s)$ is a logarithmic derivative whose solution obtained from condition:

$$\frac{\phi'}{\phi}=\frac{\pi}{\sigma} \tag{4}$$

while the function $\pi\ (s)$ and the parameter λ are defined as:

$$\pi=\left(\frac{\sigma'-\bar{\tau}}{2}\right)\pm\sqrt{\left(\frac{\sigma'-\bar{\tau}}{2}\right)^2-\bar{\sigma}+k\sigma} \tag{5}$$

$$\lambda=k+\pi' \tag{6}$$

The value of k in equation (5) can be found from the condition that the expression under the square root of equation (5) must be square of polynomial which is mostly first degree polynomial and therefore the discriminate of the quadratic expression is zero. A new eigenvalue of equation (3) is:

$$\lambda = \lambda_n = -n\pi' - \frac{n(n-1)}{2}\sigma''; \; n = 0, 1, 2, \ldots \tag{7}$$

where

$$\tau = \bar{\tau} + 2\pi \tag{8}$$

The new bound state energy is obtained using equation (6) and (7). To generate the bound state energy and the corresponding eigenfunction, the condition that $\tau' < 0$ is required. The solution of the second part of the wave function, $y_n(s)$, which is connected to Rodrigues relation [17], is given as:

$$y_n(z) = \frac{C_n}{\rho(z)} \frac{d^n}{dz^z}\{\sigma^n(z)\rho(z)\} \tag{9}$$

where C_n is normalization constant, and the weight function $\rho(s)$ must satisfies the condition:

$$\frac{\partial(\sigma\rho)}{\partial s} = \tau(s)\rho(s) \tag{10}$$

The wave function of the system is therefore obtained from equation (4), (9) and (10).

3. The Scarf Hyperbolic plus Non-Central Pocshl-Teller Potential

The Scarf Hyperbolic potential can be expressed as:

$$V = \frac{\hbar^2}{2m}\left\{\frac{b^2 + a(a+1)}{\sinh^2 r} - \frac{2b\left(a + \frac{1}{2}\right)\cosh r}{\sinh^2 r}\right\} \tag{11}$$

The visualization of this potential ($a = 2$ and $b = 2$) in 1 Dimension is:

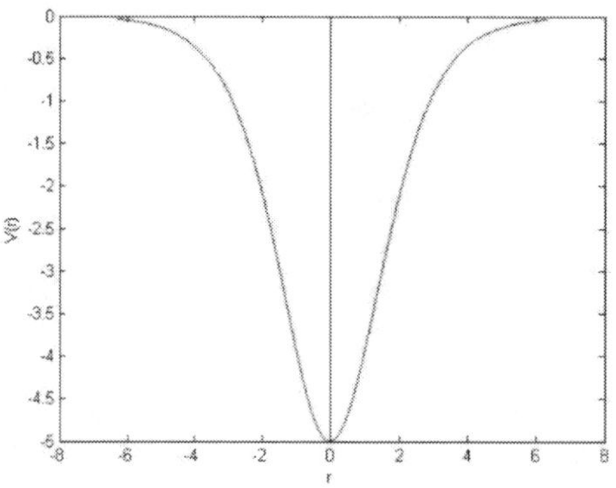

Figure 1. Visualization of Scarf Hyperbolic Potential for $a = 2$ and $b = 2$

The Scarf hyperbolic potential is symetric. This potential have a symetrical axis at $r = 0$. The Scarf potential describe particles which periodically arranged such as a crystal [11]. The application of solution in this potential are crystal model in solid state physics [5,6].

The Non Central Poschl-Teller potential can be expressed as:

$$V = \frac{\hbar^2}{2mr^2}\left\{\frac{\kappa(\kappa-1)}{\sin^2\Omega_D} + \frac{\lambda(\lambda-1)}{\cos^2\Omega_D}\right\} \tag{12}$$

The visualization of this potential ($\kappa = 2$ and $\lambda = 2$) in 1 Dimension is:

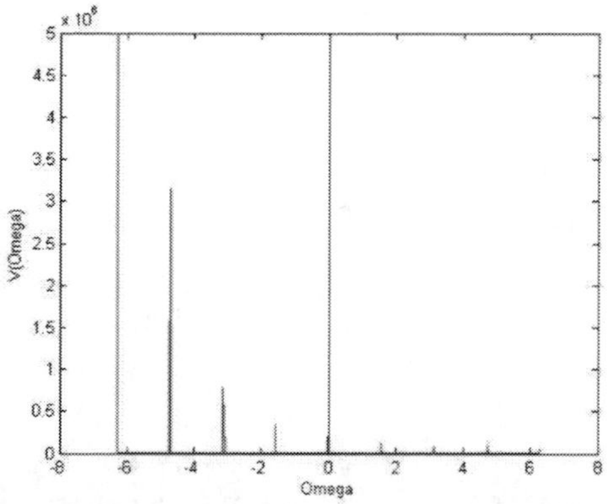

Figure 2. Visualization of Poschl-Teller Potential for $\kappa = 2$ and $\lambda = 2$

The Poschl-Teller potential like spikes which their peak decrease exponentially according to the Omega (Ω). This potential describe the particle in Poschl-Teller Oscillator. Moreover, the modified Poschl-Teller potential can been used to derive the well-known SO(2) spectrum generating algebra for an infinite square well problem [12].

The Scrodinger Equation in D-Dimensions for Scarf Hyperbolic plus Non-Central Pocshl-Teller Potential can be expressed as:

$$-\frac{\hbar^2}{2m}\nabla_D{}^2\Psi(r,\Omega)+\left[\begin{array}{l}\frac{\hbar^2}{2m}\left\{\frac{b^2+a(a+1)}{\sinh^2 r}-\frac{2b\left(a+\frac{1}{2}\right)\cosh r}{\sinh^2 r}\right\}\\ +\frac{\hbar^2}{2mr^2}\left\{\frac{\kappa(\kappa-1)}{\sin^2\Omega_D}+\frac{\lambda(\lambda-1)}{\cos^2\Omega_D}\right\}\end{array}\right]\Psi(r,\Omega)=E\Psi(r,\Omega) \tag{13}$$

where

$$\nabla_D{}^2=\frac{1}{r^{D-1}}\frac{\partial}{\partial r}\left(r^{D-1}\frac{\partial}{\partial r}\right)-\frac{\Lambda_D{}^2(\Omega_D)}{r^2} \tag{14}$$

With $\Lambda_D{}^2(\Omega_D)$ is hyperesperical harmonics, that is, D-Dimensional angular momentum operator. For $2 \leq k \leq D-1$, we have:

$$\Lambda_D{}^2(\Omega_D)=L_k{}^2=\sum_{a<b=2}^{k-1}L_{ab}{}^2=-\frac{1}{\sin^{k-1}\theta_k}\frac{\partial}{\partial\theta_k}\left(\sin^{k-1}\theta_k\frac{\partial}{\partial\theta_k}\right)+\frac{L_{k-1}{}^2}{\sin^2\theta_k} \tag{15}$$

and for $k = 1$, we have:

$$L_1{}^2=-\frac{\partial^2}{\partial\theta_1^2} \tag{16}$$

Subtitute (14), (15), (16), using variable subtitution, and separation variabel, that is,

$$\varepsilon^2=-\frac{2m}{\hbar^2}E \tag{17}$$

$$\Psi\left(r,\Omega_D=\theta_1,\theta_2,\ldots\theta_{D-1}\right)=\Psi(r)\Phi\left(\theta_1=\varphi\right)H\left(\theta_2,\ldots\theta_{D-1}\right) \tag{18}$$

we have the radial, polar, and azimuth part of Schrodinger equation are:

$$\frac{1}{r^{D-1}}\frac{\partial}{\partial r}\left(r^{D-1}\frac{\partial}{\partial r}\right)\Psi(r)-\left\{\frac{b^2+a(a+1)}{\sinh^2 r}-\frac{2b\left(a+\frac{1}{2}\right)\cosh r}{\sinh^2 r}+\varepsilon^2\right\}\Psi(r)=\frac{\lambda^{*}}{r^2}\Psi(r) \tag{19}$$

$$\frac{1}{\sin^{k-1}\theta_k}\frac{\partial}{\partial\theta_k}\left(\sin^{k-1}\theta_k\frac{\partial}{\partial\theta_k}H\right)-\frac{l_{k-1}(l_{k-1}+k-2)}{\sin^2\theta_k}H+l_k(l_k+k-1)H$$
$$-\left\{\frac{\kappa(\kappa-1)}{\sin^2\theta_k}+\frac{\lambda(\lambda-1)}{\cos^2\theta_k}\right\}H=0 \quad (20)$$

for $k = 2, 3, 4, \ldots, D-1$.

$$-\frac{1}{\Phi}\frac{d^2}{d\theta_1^2}\Phi = L_k^{\ 2} \quad (21)$$

3.1. Solution of Azimuth Part of Schrodinger Equation in D-Dimensions

The azimuth part of Schrodinger equation in D-Dimensions (equation 20) is the ordinary differential equation. The unnormalized solution of azimuth part of Schrodinger equation in D-Dimensions (21) is:

$$\Phi = Ae^{im\varphi} \quad (22)$$

where A is normalization constant.

3.2. Solution of Radial Part of Schrodinger Equation in D-Dimensions

The radial part of Schrodinger equation in D-Dimensions (equation 19) is hypergeometry differential equation. Equation (19) can be solved using some approximation and variable subtitution, that is,

$$\Psi(r)=r^{-\frac{D-1}{2}}R(r) \quad (23)$$

$$r \approx \sinh r \quad (24)$$

$$\cosh r = s \quad (25)$$

Subtitute equation (23), (24), and (25) to equation (19), we have:

$$\frac{d^2}{ds^2}R(s)+\frac{s}{(s^2-1)}\frac{d}{ds}R(s)$$
$$-\left\{\frac{b^2+a(a+1)-2b(a+\frac{1}{2})s+\varepsilon^2(s^2-1)+A_{D-1}+\frac{(D-1)(D-3)}{4}}{(s^2-1)^2}\right\}R(r)=0 \quad (26)$$

By comparing equation (1), (26), and using eigenvalue of Nikivorof-Uvarov method, we obtain the energy spectra of Scarf Hyperbolic plus Non-Central Poschl-Teller potential in D-Dimensions is:

$$E=-\frac{\hbar^2}{2m}(n-p+\tfrac{1}{2})^2 \quad (27)$$

where

$$p=\sqrt{\frac{\left\{b^2+(a+\frac{1}{2})^2+A_{D-1}+\frac{(D-1)(D-3)}{4}\right\}-\sqrt{\left[\{b+(a+\frac{1}{2})\}^2+A_{D-1}+\frac{(D-1)(D-3)}{4}\right]\left[\{b-(a+\frac{1}{2})\}^2+A_{D-1}+\frac{(D-1)(D-3)}{4}\right]}}{2}} \quad (28)$$

and A_{D-1} is centrifugal term and depend on the eigenvalue of polar part of Schrodinger equation in D-Dimensions.

Journal of Physics: Conference Series **539** (2014) 012018 doi:10.1088/1742-6596/539/1/012018

By using eigenfunction of NU method, we obtain:

$$R_n = B_n\left(s^2-1\right)^{\left(\frac{1}{4}-\frac{1}{2}p\right)}\left(s+1\right)^{-\frac{b\left(a+\frac{1}{2}\right)}{2p}}\left(s-1\right)^{\frac{b\left(a+\frac{1}{2}\right)}{2p}} P_n^{(\alpha,\beta)}(s) \tag{29}$$

where $P_n^{(\alpha,\beta)}$ is Jacobi Polynomial, that is,

$$P_n^{(\alpha,\beta)}(s) = \frac{(-1)^n}{2^n n!}(1-s)^{-\alpha}(1+s)^{-\beta}\frac{d^n}{dz^n}\left\{(1-s)^{\alpha}(1+s)^{\beta}\left(1-s^2\right)^n\right\} \tag{30}$$

and B_n is normalization constant.

Finally, from equation (23), (25), and (29), we have the radial wave function of Schrodinger equation for Scarf Hyperbolic plus Non-Central Pocshl-Teller Potentialin D-Dimensions is:

$$\Psi(r) = B_n r^{-\frac{D-1}{2}}(\sinh r)^{\frac{1}{2}-p}(\cosh r+1)^{-\frac{b}{2p}\left(a+\frac{1}{2}\right)}(\cosh r-1)^{\frac{b}{2p}\left(a+\frac{1}{2}\right)} P_n^{(\alpha,\beta)}(\cosh r) \tag{31}$$

The effect of extra dimensions on bound state energy and radial wave function can be visualized as

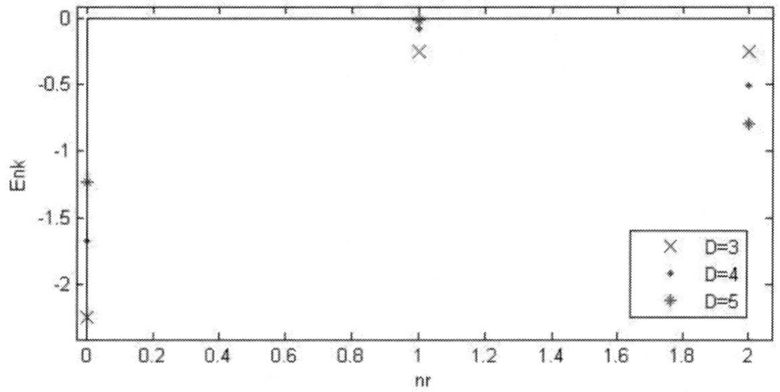

Figure 3. Bound State Energy of Scarf Hyperbolic Potential (a = 2 and b = 2)

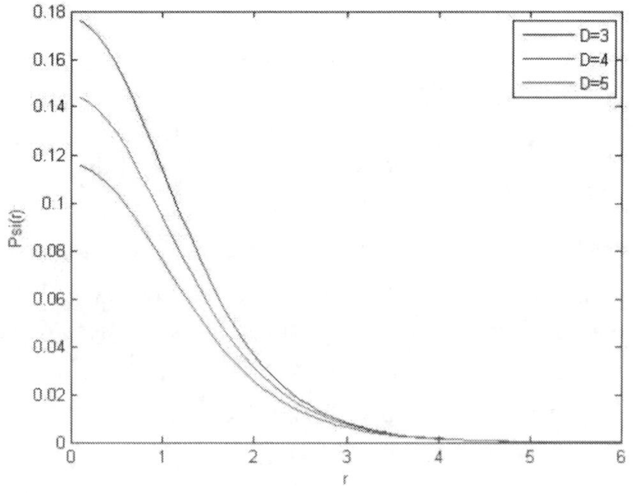

Figure 2. Radial Wave Function on Ground State for Scarf Hyperbolic Potential (a = 2 and b = 2)

From Figure 3, we have the effect of extra dimensions increase the bound state energy of system. The effect of extra dimensions in radial wave function is decrease their amplitude shown in Figure 4. It was found to agree with previous work [5,6]. Based on superstring theory [18], the amount of dimensions in the universe restricted to 10 spatial dimensions and 1 time dimension. If the amount of spatial dimension more than 10, the universe unstable and collapse. Thus, the maximum value of D is limited to 10 spatial dimensions.

3.3. Solution of Polar Part of Schrodinger Equation in D-Dimensions

According to the value of k, there are two equations at polar part of Schrodinger equation in D-Dimensions. First, for $k = 2, 3, 4, \ldots, D-1$, we solve equation (20) using variable subtitution, that is,

$$A_k = l_k \left(l_k + k - 1\right) \tag{32a}$$

$$A_{k-1} = l_{k-1} \left(l_{k-1} + k - 2\right) \tag{32b}$$

$$s = \cos 2\theta_k \tag{33}$$

Subtitute equation (32a), (32b), and (33) to (21a), we have

$$\frac{d^2}{ds^2} H - \frac{\frac{1}{2}\left[(k-1)+(k+1)s\right]}{\left(1-s^2\right)} \frac{d}{ds} H - \left\{ \frac{\frac{1}{2}\kappa(\kappa-1)(1+s)+\frac{1}{2}\lambda(\lambda-1)(1-s)+\frac{1}{2}A_{k-1}(1+s)-\frac{1}{4}A_k\left(1-s^2\right)}{\left(1-s^2\right)^2} \right\} H = 0 \tag{34}$$

By comparing equation (1), (34), andusing eigenvalue of Nikivorof-Uvarov method, we obtain the eigenvalue of Scarf Hyperbolic plus Non-Central Pocshl-Teller potential in D-Dimensions:

$$A_k = 4\left(n+p+\tfrac{1}{2}\right)^2 - \tfrac{1}{4}(k-3)^2 - (k-2) \tag{35}$$

where

$$p = \sqrt{\tfrac{1}{2}(q+t)+\tfrac{1}{2}\sqrt{4qt-(k-2)^4-2(k-2)^2(q-t)}} \tag{36}$$

$$q = \tfrac{1}{2}\left[\kappa(\kappa-1)+A_{k-1}\right] \tag{37}$$

$$t = \tfrac{1}{2}\lambda(\lambda-1)+\tfrac{1}{16}(k-1)^2+\tfrac{1}{16}(k-3)^2 \tag{38}$$

By using eigenfunction of NU method, we obtain:

$$H\left(s_k\right) = B_n \left(1-s^2\right)^{-\frac{1}{2}\left(\frac{1}{4}(k-3)-p\right)} (1+s)^{\frac{1}{2}\left\{\frac{1}{4}(k-1)-\frac{1}{2p}\left[(k-2)^2+q-t\right]\right\}} (1-s)^{-\frac{1}{2}\left\{\frac{1}{4}(k-1)-\frac{1}{2p}\left[(k-2)^2+q-t\right]\right\}} P_n^{(\alpha,\beta)}(s) \tag{39}$$

Finally, from equation (33), (36), (37), (38), and (39), we have the first equation ofpolar wave function Schrodinger equation for Scarf Hyperbolic plus Non-Central Pocshl-Teller Potentialin D-Dimensions ($k = 2, 3, 4, \ldots, D-1$), that is:

$$H\left(\cos 2\theta_k\right) = B_{n_{\theta_k}} \left(\sin 2\theta_k\right)^{-\frac{1}{2}\left(\frac{1}{4}(k-3)-p_{\theta_k}\right)} \left(\cos\theta_k\right)^{\left\{\frac{1}{4}(k-1)-\frac{1}{2p_{\theta_k}}\left[(k-2)^2+q_{\theta_k}-t_{\theta_k}\right]\right\}} \left(\sin\theta_k\right)^{-\left\{\frac{1}{4}(k-1)-\frac{1}{2p_{\theta_k}}\left[(k-2)^2+q_{\theta_k}-t_{\theta_k}\right]\right\}} P_{n_{\theta_k}}^{\left(\alpha_{\theta_k},\beta_{\theta_k}\right)}\left(\cos 2\theta_k\right) \tag{40}$$

4. Special Case for the Ground and First Excited State in 3 Dimensions

Special case for 3 Dimensional problem for Scarf Hyperbolic plus Non-Central Pocshl-Teller Potential, that is:

$$\Psi\left(r,\Omega_3\right) = \Psi(r) Y\left(\Omega_3\right) \tag{41}$$

with solution of radial part of 3 dimensional Schrodinger equation:

$$\Psi(r)=B_{n_r} r^{-1}(\sinh r)^{\frac{1}{2}-p_r}(\cosh r+1)^{-\frac{b}{2p_r}\left(a+\frac{1}{2}\right)}(\cosh r-1)^{\frac{b}{2p_r}\left(a+\frac{1}{2}\right)} P_{n_r}^{(\alpha_r,\beta_r)}(\cosh r) \quad (42)$$

$$B_{n_r}=(-1)^{\frac{b\left(a+\frac{1}{2}\right)}{2p_r}-\frac{1}{2}p_r-\frac{1}{2}}\sqrt{\frac{2n+\alpha_r+\beta_r+1}{2^{\alpha_r+\beta+1}}\frac{\Gamma(n_r+\alpha_r+\beta_r+1)n_r!}{\Gamma(n_r+\alpha_r+1)\Gamma(n_r+\beta_r+1)}} \quad (43)$$

$$P_{n_r}^{(\alpha_r,\beta_r)}(\cosh r)=\frac{(-1)^{n_r}}{2^n n_r!}(1-\cosh r)^{-\alpha_r}(1+\cosh r)^{-\beta_r}\frac{d^{n_r}}{d(\cosh r)^{n_r}}\left\{\begin{matrix}(1-\cosh r)^{\alpha_r}(1+\cosh r)^{\beta_r}\\(\sinh^2 r)^{n_r}\end{matrix}\right\} \quad (44)$$

$$\alpha_r=\frac{b\left(a+\frac{1}{2}\right)}{p_r}-p_r \quad (45)$$

$$\beta_r=-\frac{b\left(a+\frac{1}{2}\right)}{p_r}-p_r \quad (46)$$

$$p_r=\sqrt{\frac{\left\{b^2+\left(a+\frac{1}{2}\right)^2+l(l+1)\right\}-\sqrt{\begin{bmatrix}\left\{b+\left(a+\frac{1}{2}\right)\right\}^2+l(l+1)\\\left\{b-\left(a+\frac{1}{2}\right)\right\}^2+l(l+1)\end{bmatrix}}}{2}} \quad (47)$$

From equation (42) to (47), we can make a visualization of radial wave function with Matlab, that is:

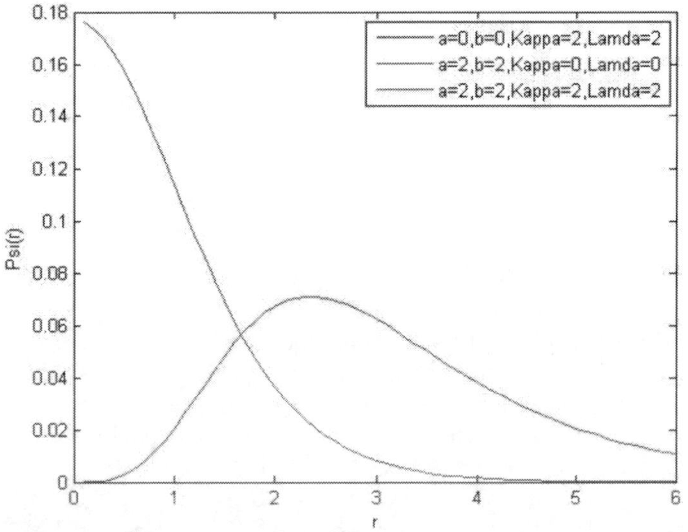

Figure 5. Radial Ground State Wave Function in 3 Dimension for Scarf Hyperbolic plus Non-Sentral Poschl-Teller Potential

From Figure 5, we have the effect of each potential parameters. The effect of parameter Non Central Poschl-Teller won't appear if the parameters of Scarf hyperbolic is zero. It causes the radial wave function vanish. If the Scarf hyperbolic parameters not zero, the Non Central Poschl-Teller parameters decrease the radial wave function and move wave function to right.

Journal of Physics: Conference Series **539** (2014) 012018 doi:10.1088/1742-6596/539/1/012018

Solution of Angular part devided into 2 parts, that is:

$$Y_{l_1}^{(l)}\left(\Omega_D=\theta_1,\theta_2\right)=\Phi\left(\theta_1=\varphi\right)H\left(\theta_2\right) \tag{48}$$

Solution of azimuth part equation:

$$\Phi = Ae^{im\varphi} \tag{49}$$

$$A=\frac{1}{\sqrt{2\pi}} \tag{50}$$

Solution of polar part equation:

$$H\left(\cos 2\theta_2\right)=B_n\left(1+\cos 2\theta_2\right)^{\frac{\sqrt{t_{\theta_2}}}{\sqrt{2}}+\frac{1}{4}}\left(1-\cos 2\theta_2\right)^{\frac{\sqrt{q_{\theta_2}}}{\sqrt{2}}}P_n^{(\alpha,\beta)}\left(\cos 2\theta_2\right) \tag{51}$$

with

$$P_{n_{\theta_2}}^{\left(\alpha_{\theta_2},\beta_{\theta_2}\right)}\left(\cos 2\theta_2\right)=\frac{\left(-1\right)^{n_{\theta_2}}}{2^{n_{\theta_2}+\alpha_{\theta_2}+\beta_{\theta_2}}n_{\theta_2}!}\left(\sin^2\theta_2\right)^{-\alpha_{\theta_2}}\left(\cos^2\theta_2\right)^{-\beta_{\theta_2}}$$
$$\frac{d^{n_{\theta_2}}}{d\left(\cos 2\theta_2\right)^{n_{\theta_2}}}\left\{\left(\sin^2\theta_2\right)^{\alpha_{\theta_2}+n_{\theta_2}}\left(\cos^2\theta_2\right)^{\beta_{\theta_2}+n_{\theta_2}}\right\} \tag{52}$$

$$p_{\theta_2}=\tfrac{1}{\sqrt{2}}\left(\sqrt{q_{\theta_2}}+\sqrt{t_{\theta_2}}\right) \tag{53}$$

$$q_{\theta_2}=\tfrac{1}{2}\left[\kappa\left(\kappa-1\right)+m^2\right] \tag{54}$$

$$t_{\theta_k}=\tfrac{1}{2}\lambda\left(\lambda-1\right)+\tfrac{1}{8} \tag{55}$$

$$\alpha=\sqrt{2t_{\theta_2}} \tag{56}$$

$$\beta=\sqrt{2q_{\theta_2}} \tag{57}$$

$$l=2\left(n_{\theta_k}+p_{\theta_k}+\tfrac{1}{4}\right) \tag{58}$$

Using subtitution of all equation (48) to (58), we get unnormalized solution of 3 dimensional Schrodinger equation for Scarf hiperbolik plus non sentral Poschl-Teller potential, for example:

Table 1. Visualization of angular part wave function for Non-Sentral Poschl-TellerPotential

No	$H\left(\cos 2\theta\right)_{n_l m\kappa\lambda}$	3 Dimension	2 Dimensional Projection at x-z axis
1	$H\left(\cos 2\theta_2\right)_{0000}$	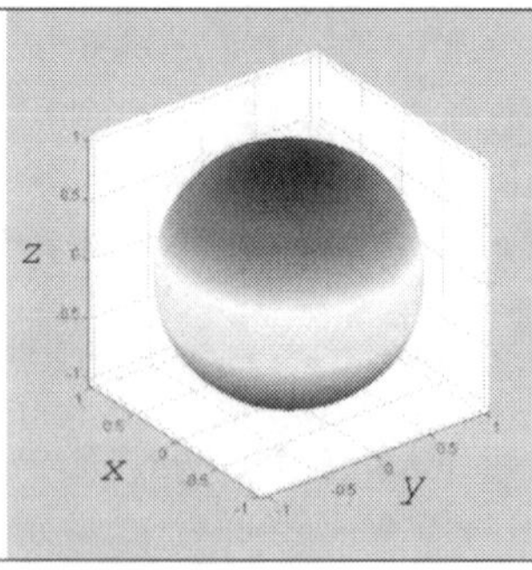	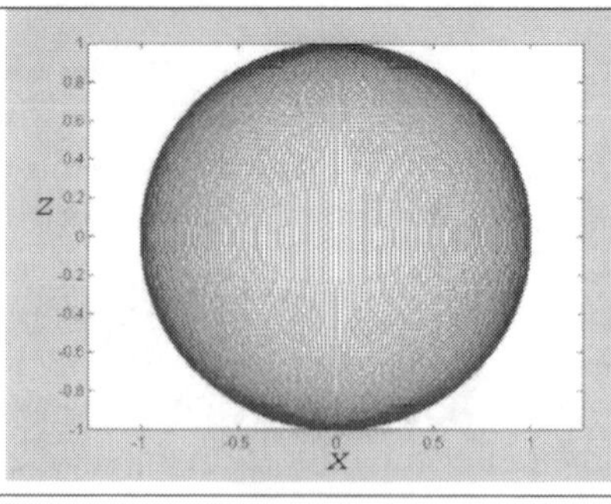

No	$H(\cos 2\theta)_{n_l m \kappa \lambda}$	3 Dimension	2 Dimensional Projection at *x-z* axis
2	$H(\cos 2\theta)_{0033}$	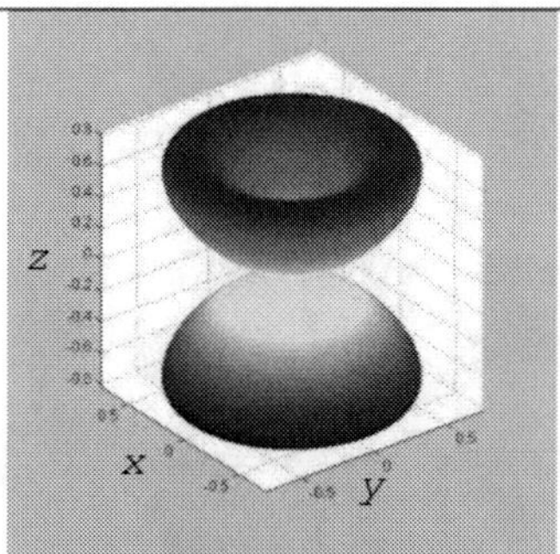	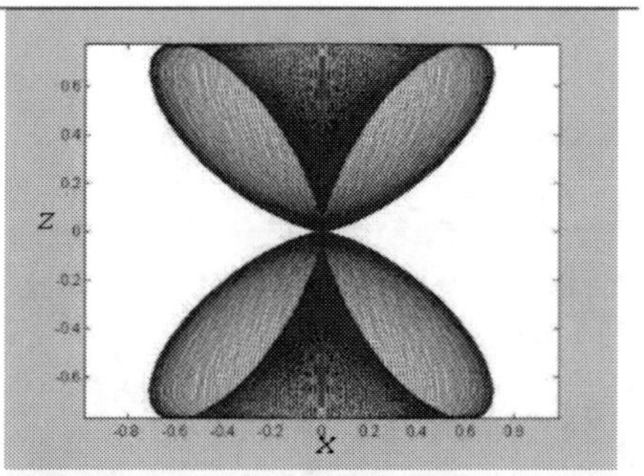
3	$H(\cos 2\theta)_{0100}$	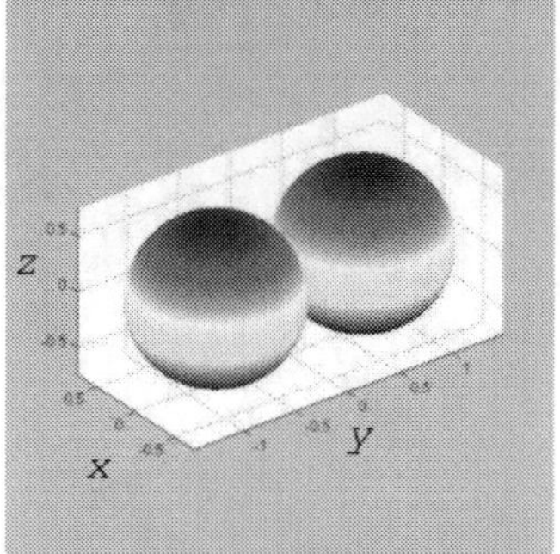	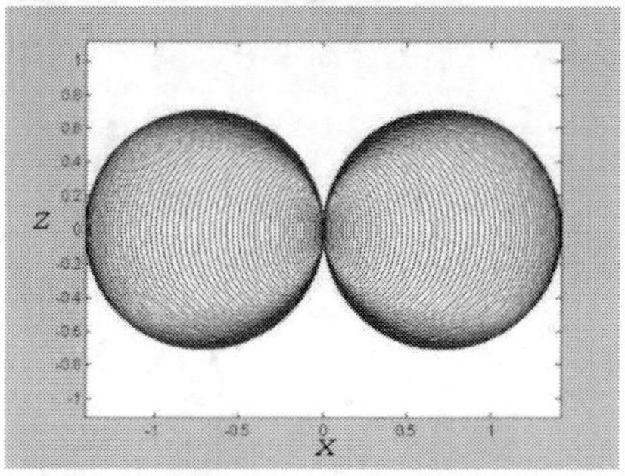
4	$H(\cos 2\theta)_{0133}$	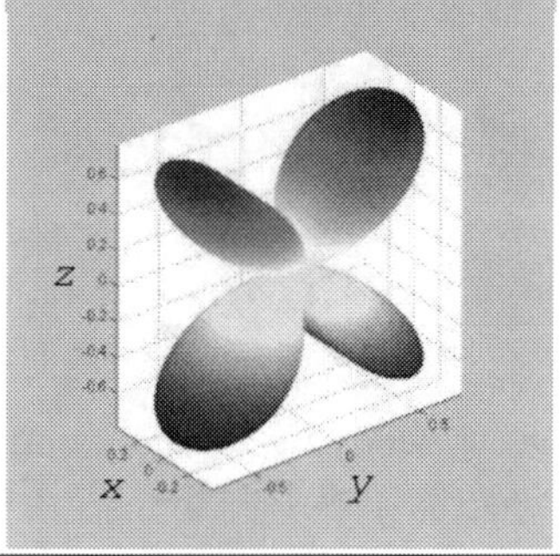	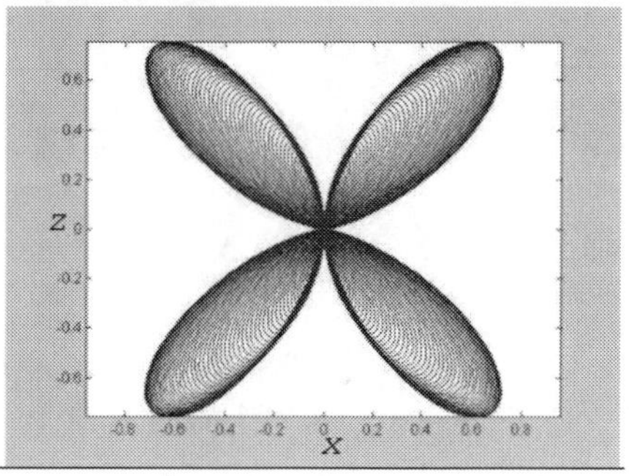

The effect of polar potential parameters in 3-dimensions Non-Central Poschl-Teller potential on orbital (sub shell) electrons can be described in Table 1. Without polar potential (κ and λ is zero), the orbital electron will be changed into orbitals in spherical harmonics or hydrogen-like atom. This phenomena figured in the number 1and 3, which corresponds to an orbital in spherical harmonics in the same parameters [19].

The effect of wave function parameter (λ) make the wave function pulled to z axis and having a reflection of the x-y plane so that the wave function looks like 2 pieces of adjacent balloons. The effect of κ parameter rotate wave function with direction φ with the rotary axis at the origin so that the wave function looks like a donut. When these two parameters given, the effect of each parameter affects the wave function. The combination of these effects result in Table 3 number 2 and 4.

The energy spectra of 3 dimensional Schrodinger equation for Scarf hyperbolic plus non central Poschl-Teller potential, that is:

$$E = -\frac{\hbar^2}{2m}\left(n_r - p_r + \tfrac{1}{2}\right)^2 \tag{59}$$

with

$$p_r = \sqrt{\frac{\left\{b^2+\left(a+\frac{1}{2}\right)^2+l(l+1)\right\}-\sqrt{\left[\left\{b+\left(a+\frac{1}{2}\right)\right\}^2+l(l+1)\right]\left[\left\{b-\left(a+\frac{1}{2}\right)\right\}^2+l(l+1)\right]}}{2}} \tag{60}$$

$$l(l+1)=4\left(n_{\theta_2}+p_{\theta_2}+\tfrac{1}{2}\right)^2-\tfrac{1}{4} \tag{61}$$

$$p_{\theta_2}=\tfrac{1}{\sqrt{2}}\left(\sqrt{q_{\theta_2}}+\sqrt{t_{\theta_2}}\right) \tag{62}$$

$$q_{\theta_2}=\tfrac{1}{2}\left[\kappa(\kappa-1)+m^2\right] \tag{63}$$

$$t_{\theta_2}=\tfrac{1}{2}\lambda(\lambda-1)+\tfrac{1}{8} \tag{64}$$

From equation (59) to (64), we can make a visualization of bound state energy with Matlab, that is:

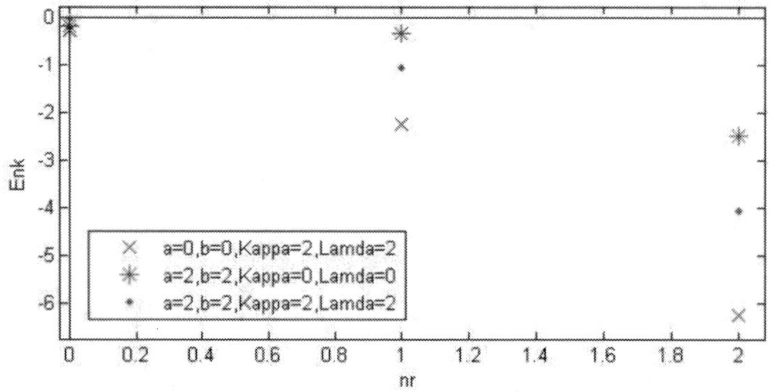

Figure 6 Bound State Energy in 3 Dimension for Scarf Hyperbolic plus Non-Sentral Poschl-Teller Potential

From figure 6, we have the bound state energy for each value of potential parameters. The effect of Scarf Hyperbolic potential increases the bound state energy of system. In the other hand, the effect of non central Poschl-Teller potential decreases the bound state energy of system

5. Concluding Remarks

In this paper, we present the solutions of Schrodinger equation in D-dimension for Scarf hyperbolic plus non central Poschl-Teller potential within the framework of an approximation to the centrifugal and high dimension term. The bound state energy were obtained in D-dimensions using the Nikiforov-Uvarov method, and it was found to agree with previous works [5,10-11]. The corresponding wave function of the Scarf hyperbolic plus non central Poschl-Teller potential were obtained in terms of the Jacobi polynomials. The example of bound state energy and wave function in 3 dimensions presented in condition of ground state. The existence of arbitrary dimensions increase bound state energy system. In the other hand, the existence of arbitrary dimensions decreases the amplitude of wave function. The effect of Scarf Hyperbolic potential increases the bound state energy of system. The effect of non central Poschl-Teller potential decreases the bound state energy of system.

Acknowledgement
This work is partly funded by Hibah Pascasarjana DIKTI, Sebelas Maret University (UNS), No.165/UN27.11/PN2013.

Journal of Physics: Conference Series **539** (2014) 012018 doi:10.1088/1742-6596/539/1/012018

References

[1] Chen G 2004 *Z. Naturforsch* **59a** 875 – 876
[2] Dong SH and Sun GH 2003 *Physics Letters A* **314** 261–266
[3] Agboola D 2010 *Chinese Physics Letters* **27** 040301
[4] Agboola D 2009 *Physica Scripta* **80** 065304
[5] Deta UA, Suparmi, and Cari 2013 *Adv. Studies Theor. Phys.* **7/13** 647–656
[6] Deta UA, Suparmi, and Cari 2013 *AIP Conf. Proc.* **1554** 190
[7] Greene RI and Aldrich C 1976 *Phys. Rev. A* **14** 2363
[8] Xu Y, He S, and Jia CS 2010 *Phys. Scr.* **81** 045001
[9] Nikivorof AF and Uvarov VB 1988 *Special Functions of Mathematical. Physics – A Unified Introduction with Applications* (Germany: Birkhauser Verlag Basel) pp. 317-318
[10] Negro J, Nieto LM, and Ortiz OR 2004 *Journal of Physics A: Mathematical and General* **37/43** 10079.
[11] Castillo 2007 *Revista Mexicana de Fisica* **E53 2** 143–154.
[12] Dong SH, Lemus R *A New Dynamical Group Approach to the Modified Poschl-Teller Potential* arXiv: quant-ph/0110157
[13] Suparmi 2011 *Mekanika Kuantum II* (Surakarta: Physics Department-Faculty of MIPA-Sebelas Maret University) pp. 166-169
[14] Ikot AN, Akpabio LE, and Uwah E J 2011 *EJTP* **8/25** 225-232
[15] Ikot AN, Antia AD, Akpabio LE, and Obu JA 2011 *JVR* **6/2** 65-76
[16] Inomata A , Suparmi A, and Kurth S 1991 *Proceeding of 18th International Colloquium on Group Theoretical Methods in Physics* V 399
[17] Yasuk F, Berkdemir C, and Berkdemir R 2005 *J. Phys. A: Math. Gen* **38** 65795
[18] Wray K 2011 *An Introduction to String Theory* (Online) http://math.berkeley.edu/~kwray/papers/string_theory.pdf accessed 11 Juli 2013
[19] Orchin M, Macomber RS, Pinhas AR, Wilson RM 2005 *The Vocabulary and Concepts of Organic Chemistry - Second Edition* (John Wiley & Sons, Inc., Hoboken, NJ, USA) pp 13-16

CTPNP 2014: "From Universe to String's Scale" IOP Publishing
Journal of Physics: Conference Series **539** (2014) 012019 doi:10.1088/1742-6596/539/1/012019

Analysis of Energy and Wave Function For Manning-Rosen Plus Scarf Potential D-Dimension With Nikiforov Uvarov Method

Luqman H[1], Suparmi [2], Cari[2] and U.A. Deta[3]

[1]Physics Education Program, Palangkaraya University, Indonesia
Jl. Yos Sudarso, Palangkaraya 7311A, Indonesia

[2]Physics Department of Graduate Program
Sebelas Maret University, Indonesia
Jl. Ir. Sutami 36A Kentingan, Surakarta 57126, Indonesia

[3]Department of Physics, Faculty of Mathematics and Natural Science
State University of Surabaya, Indonesia
Jl. Ketintang, Surabaya 60231, Indonesia

Email: luqman_812_h@yahoo.com

Abstract. Approximate analysis of D-dimensional Schrödinger equation for Manning Rosen Plus Scarf potential investigated by Nikiforov-Uvarov method. The approximate energy of Manning Rosen plus Scarf is expressed of closed form. The approximate wave function is expressed of general Jacobi polynomial.

1. Introduction

Quantum mechanics always use a different approach to determine magnitude relevant with the motion of the particles is to use the wave function for the dynamics of particles moving represent to obtained from solution of the Schrödinger equation relating particle. In quantum mechanics, exactly solutions of the wave equation with certain physical potentials were of great interest. Schroodinger equation is a partial differential equation that describes the quantum state of a physical system. Solution of the Schrödinger equation is exact on some potentially be applied in a physical system that is very important because it can be used to determine the wave functions and energy levels of a system of particles.

Several methods can be used to solve the equations Schroodinger, among other SUSY [3], WKB approach [4], Romanovsky [5], and Nikiforov-Uvarov [6]. Describe the potential in the quantum dynamics of particles in quantum mechanics. Some examples of potential in quantum mechanics, among others, potential Coloumb, Rosen Morse, Manning-Rosen, Scarf, Poschl Teller Eckart and others.

In this work used the NU method to obtain the energy and the wave function of the non-central Manning Rosen potential plus Scarf. NU method has been used to complete the analysis of some of potential, such of Scarf Hiperbolik potential, Deng Fan deform plus Eckart deform potential, Klatzer deform plus Ring Shaped deform potential.

2. Schrodinger equation in D-Dimension

Schroodinger equations in D dimension based on the use of D-dimensional polar coordinate with polar variable r (hyper radius) and the angular momentum variables $\theta_1, \theta_2, \theta_3 \ldots\ldots, \theta_{D-1}, \phi$(hyper angle). For D-dimensional space, the Schrodinger equation can be write :

$$-\frac{\hbar^2}{2m}\nabla_D^2\Psi(r,\Omega_D) + V(r,\Omega_D)\Psi(r,\Omega_D) = E\Psi(r,\Omega_D) \quad (1)$$

Where the Laplacian operator is $\nabla_D^2 = \frac{1}{r^{D-1}}\frac{\partial}{\partial r}\left(r^{D-1}\frac{\partial}{\partial r}\right) - \frac{\Lambda_D^2(\Omega_D)}{r^2}$ (2)

The second term on the right-hand side of Eq. (2) is the multidimensional space centrifugal term, and Ω_D represents the angular coordinates. In this case, the operator Λ_D^2 yields a hyperspherical harmonic as its eigenfunction, mathematically

$$\Lambda_D^2(\Omega_D) = L_k^2 = \sum_{a<b=2}^{k-1} L_{ab}^2 = -\frac{1}{\sin^{k-1}\theta_k}\left(\sin^{k-1}\theta_k\frac{\partial}{\partial\theta_k}\right) + \frac{L_{D-1}^2}{\sin^2\theta_k} \quad (3)$$

The eq.(2) helps us to write the wave function as

$$\Psi_{nlm}(r,\Omega_D) = R_{n,l}(r)\, Y_l^m(\Omega_D) \quad (4)$$

where $R_{n,l}(r)$ is radial part of equation and $Y_l^m(\Omega_D)$ is the anguler part called hypersperical harmonics. The $Y_l^m(\Omega_D)$ obeys the eigenvalue equation

$$\Lambda_D^2\, Y_l^m(\Omega_D) = l(l+D-2)\, Y_l^m(\Omega_D) \quad (5)$$

3. Nikiforov-Uvarov (NU) Method

Nikiforov Uvarov method is based on the sulution of a second order differential equation by using an intermediary hipergeometry equation. The Schrodinger equation $\Psi''(r) + [E - V(r)]\Psi(r) = 0$ (6) can be solved by transforming it into a hypergeometric-type equation through using the transformation, s = s(x) and its resulting equation is expressed as

$$\psi''(s) + \frac{\tilde{\tau}(s)}{\sigma(s)}\psi'(s) + \frac{\tilde{\sigma}(s)}{\sigma^2(s)}\psi(s) = 0 \quad (7)$$

where $\sigma(s)$ and $\tilde{\sigma}(s)$ are second order polynomial, $\tilde{\tau}(s)$ is first order polynomial, and $\psi(s)$ is a function of hypergeometry. To solve eq.(7), we set the wave function as

$$\Psi(s) = \varphi(s)\chi_n(s) \quad (8)$$

Subsituting eq.(8) into eq.(7) reduce eq.(7) into a hypergeometric type equation

$$\sigma(s)\chi_n''(s) + \tau(s)\chi_n'(s) + \lambda\chi_n(s) = 0 \quad (9)$$

where the wave function $\varphi(s)$ is defined as a logarithmic derivative. The first part of the wave function can be describe as

$$\frac{\phi'}{\phi} = \frac{\pi}{\sigma} \quad (10)$$

and the other wave function $\chi_n(s)$ is the hypergeometric-type function whose polynomials are given by the Rodrigues relation

$$\chi_n(s) = \frac{B_n}{\rho(s)}\frac{d^n}{ds^n}[\sigma^n(s)\rho(s)] \quad (11)$$

where B_n is the normalization constant and weight function $\rho(s)$ satisfies the Pearson equation is expressed as:

$$\frac{d}{ds}(\sigma(s)\rho(s)) = \tau(s)\rho(s) \quad (12)$$

The function $\pi(s)$ and the parameter λ required for NU method are defined as follows:

$$\pi = \left(\frac{\sigma' - \tilde{\tau}}{2}\right) \pm \sqrt{\left(\frac{\sigma' - \tilde{\tau}}{2}\right)^2 - \tilde{\sigma} + k\sigma} \quad (13)$$

$$\lambda = k + \pi' \tag{14}$$

It is necessary that the term under the square root sign in Eq. (13) is the square of a polynomial. To calculate k from Eq. (13), the discriminant of the quadratic term must vanish. The eigenvlues in Eq. (14) take the form

$$\lambda = \lambda_n = -n\tau' - \frac{n(n-1)}{2}\sigma'', \quad n = 0,1,2,\ldots \tag{15}$$

Where

$$\tau(s) = \bar{\tau}(s) + 2\pi(s) \tag{16}$$

and its derivative is less than zero, which is the necessary condition for bound state solutions. The energy eigenvalues are obtained by comparing Eq. (14) with Eq. (15).

4. Analysis Manning Rosen Plus Scarf Potential D dimension with Nikiforov-Uvarov Method

The non-central potential which is combination of Manning Rosen potential and Scarf non-cental potential given as

$$V_{(r,\Omega_D)} = \frac{\hbar^2}{2m}\left(\frac{v(v-1)}{\sinh^2 r} - 2q\coth r\right) + \frac{\hbar^2}{2mr^2}\left(\frac{b^2 + a(a-1)}{\sin^2\Omega_D} - \frac{2b\left(a-\frac{1}{2}\right)\cos\Omega_D}{\sin^2\Omega_D}\right) \tag{17}$$

where $b^2 + a(a-1) > 0$ and $2b\left(a-\frac{1}{2}\right) > 0$.

Subtitution of Manning Rosen plus Scarf potential in Schrodinger equation D dimension, then separation of variables to obtained

$$\frac{1}{R(r)\,r^{D-1}}\frac{\partial}{\partial r}\left(r^{D-1}\frac{\partial R(r)}{\partial r}\right)r^2 - \left(\frac{v(v-1)}{\sinh^2 r} - 2q\coth r + \varepsilon^2\right)r^2 = l(l+D-2) \tag{18}$$

$$\frac{1}{Y(\Omega_D)}\,\Lambda_D^2(\Omega_D)Y(\Omega_D) + \left(\frac{b^2+a(a-1)}{\sin^2\Omega_D} - \frac{2b\left(a-\frac{1}{2}\right)\cos\Omega_D}{\sin^2\Omega_D}\right) = l(l+D-2) \tag{19}$$

Eq (18) and eq(19) are radial part and polar part respectively.

Solution of the radial part in equation (18), performed by intermediaries hypergeometry equation can write as $\frac{d^2\chi(r)}{dr^2} - \left[\frac{\left(l+\frac{D-1}{2}\right)\left(l+\frac{D-3}{2}\right)}{r^2}\right]\chi(r) - \left(\frac{v(v-1)}{\sinh^2 r} - 2q\coth r + \varepsilon^2\right)\chi(r) = 0$ (20)

If $\coth r = s$ and $\frac{1}{r^2} = \left(d_0 + \frac{1}{\sinh^2 r}\right)$ the equation (20) can rewritten as

$$\frac{d^2\chi(s)}{ds^2} - \frac{2s}{(1-s^2)}\frac{d\chi(s)}{ds} - \left[\frac{\left(l+\frac{D-1}{2}\right)\left(l+\frac{D-3}{2}\right)d_0}{(1-s^2)^2} - \frac{\left(l+\frac{D-1}{2}\right)\left(l+\frac{D-3}{2}\right)(1-s^2)}{(1-s^2)^2} - \frac{v(v-1)(1-s^2)}{(1-s^2)^2} - \frac{2qs}{(1-s^2)^2} + \frac{\varepsilon^2}{(1-s^2)^2}\right]\chi(s) = 0 \tag{21}$$

Equation (21) is differential equations that can be solved by Nikiforov-Uvarov method. By substituting each of parameter NU, we obtained:

$$\pi = \pm\sqrt{\begin{aligned}&\left(\left(l+\frac{D-1}{2}\right)\left(l+\frac{D-3}{2}\right)d_0 + v(v-1) - k\right)s^2 - 2qs\\ &\quad + \left(l+\frac{D-1}{2}\right)\left(l+\frac{D-3}{2}\right)d_0 + \varepsilon^2\\ &\quad - \left(\left(l+\frac{D-1}{2}\right)\left(l+\frac{D-3}{2}\right)d_0 + v(v-1) - k\right)\end{aligned}} \tag{22}$$

Energy eigen values and eigen functions are

$$\lambda_1 = \lambda_2 = \left(l+\frac{D-1}{2}\right)\left(l+\frac{D-3}{2}\right)d_0 + v(v-1) - p^2 - p \tag{23}$$

and new eigen value is $\lambda_n = 2n(p+1) + n(n+1) = 2np + n^2 + n$ (24)

$$\text{where } p = \sqrt{\left(l+\frac{D-1}{2}\right)\left(l+\frac{D-3}{2}\right)d_0 + \left(v-\frac{1}{2}\right)^2} - \left(n+\frac{1}{2}\right) \tag{25}$$

From eq(19), separation variable azimuth and polar, we obtained

$$\Phi = \frac{1}{\sqrt{2\pi}} e^{im\varphi} \tag{26}$$

$$\text{and } \frac{\partial^2 \mathrm{H}}{\partial \theta_k^2} + (k-1)\cot\theta_k \frac{\partial \mathrm{H}}{\partial \theta_k} - \left(\frac{A_{k-1}}{\sin^2\theta_k} - A_k\right)\mathrm{H} - \left(\frac{b^2+a(a-1)}{\sin^2\theta_k} - \frac{2b\left(a-\frac{1}{2}\right)\cos\theta_k}{\sin^2\theta_k}\right)\mathrm{H} = 0 \tag{27}$$

where eq.(26) and eq(27) are solution of azimuth part and polar part equation.
To solving eq.(27),we substituted if $\cos\theta_k = s$, then eq.(27) can rewritten as

$$\frac{d^2}{ds^2}H - \frac{ks}{(1-s^2)}\frac{d}{ds}H - \left\{\frac{b^2+a(a-1)}{(1-s^2)^2} - \frac{2b\left(a-\frac{1}{2}\right)s}{(1-s^2)^2} + \frac{A_{k-1}}{(1-s^2)^2} - \frac{A_k(1-s^2)}{(1-s^2)^2}\right\}H = 0 \tag{28}$$

Equation (28) is differential equations that can be solved by Nikiforov-Uvarov method.

By substituting each of parameter NU, we obtained:

$$\pi = \left(\frac{(k-2)s}{2}\right) \pm \sqrt{\begin{array}{c}\left(\frac{(k-2)^2}{4} + A_k - k^*\right)s^2 - 2b\left(a-\frac{1}{2}\right)s + b^2 + a(a-1) \\ +A_{k-1} - A_k + k^* - \frac{(k-2)^2}{4} + \frac{(k-2)^2}{4}\end{array}} \tag{29}$$

Energy eigen values and eigen functions are

$$\lambda_1 = \lambda_2 = \frac{(k-2)^2}{4} + A_k - p^2 + p - \frac{(k-2)}{2} \tag{30}$$

and new eigen value is

$$\lambda_n = 2np + 2n + n(n-1) = 2np + n^2 + n \tag{31}$$

where

$$p = \sqrt{\frac{\left(b^2+a(a-1)+A_{k-1}+\frac{(k-2)^2}{4}\right) + \sqrt{\left(b^2+a(a-1)+A_{k-1}+\frac{(k-2)^2}{4}\right)^2 - 4b^2\left(a-\frac{1}{2}\right)^2}}{2}} \tag{32}$$

The energy of Manning Rosen plus Scarf potential D dimension is

$$E_n = -\frac{\hbar^2}{2m}\left[\left\{\sqrt{\left(l+\frac{D-1}{2}\right)\left(l+\frac{D-3}{2}\right)d_0 + \left(v-\frac{1}{2}\right)^2} - \left(n+\frac{1}{2}\right)\right\}^2 + \frac{q^2}{\left\{\sqrt{\left(l+\frac{D-1}{2}\right)\left(l+\frac{D-3}{2}\right)d_0 + \left(v-\frac{1}{2}\right)^2} - \left(n+\frac{1}{2}\right)\right\}^2} - \left(l+\frac{D-1}{2}\right)\left(l+\frac{D-3}{2}\right)d_0\right] \tag{33}$$

In the special condition, where $(l = 0)$ and $(D = 3)$, we obtained

$$E_n = -\frac{\hbar^2}{2m}\left\{(v-1-n)^2 + \frac{q^2}{(v-1-n)^2}\right\} \tag{34}$$

Angular momentum parameter of Manning Rosen plus Scarf potential D dimension in the general condition obtained by $\lambda_1 = \lambda_{n_l}$ and we obtained a relationship that

$$A_k = \left((n_l + p) + \frac{1}{2}\right)^2 - \left(\frac{(k-2)}{2} + \frac{1}{2}\right)^2 \tag{35}$$

The wave function expressed by generally Jacobi polynomial. The wave function of radial part is

$$R_n = B_n\, r^{-\left(\frac{D-1}{2}\right)}(1 - coth^2 r)^{\frac{p}{2}}(1 + \coth r)^{\frac{q}{2p}}(1 - \coth r)^{-\frac{q}{2p}} P_n^{\alpha,\beta}(\coth r) \tag{36}$$

and the wave function of polar part is

$$H = B_n\,(1 + \cos\theta)^{\frac{b\left(a-\frac{1}{2}\right)}{2p} - \frac{(k-p-2)}{2}}(1 - \cos\theta)^{-\frac{b\left(a-\frac{1}{2}\right)}{2p} - \frac{(k-p-2)}{2}} P_n^{\alpha,\beta}(\cos\theta) \tag{37}$$

5. Conclusion

The In this work we present the approximate solution of Schrodinger equation for Manning Rosen plus Scarf potential in D dimension with Nikiforov Uvarov method. The energy of Manning Rossen plus Scarf obtained by equating the eigenvalues. The wave function expressed by generally Jacobi polynomial.

Acknowledgement

This work is partly funded by Hibah Pascasarjana DIKTI, Sebelas Maret University (UNS), No.165/UN27.11/PN2013.

References

[1] AD Antia, AN Ikot, and LE Akpabio 2010 *European Journal of Scientific Research* **46** 107-118.
[2] A Awoga, AN Ikot, and B Emah 2013 Reseach Revista Mexicana de Fısica 59 229–235.
[3] UA Deta, Suparmi, Cari 2013 Adv. Studies Theor. Phys. 7/13 647-656.
[4] SM Ikhdair and R Sever 2007 arxiv.org 1-16.
[5] Cari and Suparmi 2012 *International Journal of Applied Physics and Mathematics* **2/3**.
[6] DJ Griffiths 1995 *Introduction to Quantum Mechanics* (New Jersey: Prentice Hall)
[7] H Hassanabadi, S Zarrinkamar, and Rajabi 2011 *Commun. Theor. Phys.* **55** 541–544.
[8] AN Ikot, A Awoga,and A Antias 2013 *Chin. Phys. B* **22/2** 020304.
[9] S Meyur and S Debnath 2009 *Lat. Am. J. Phys. Educ.* **3/2**.
[10] AV Nikiforov and VB Uvarov 1988 *Special Functions of Mathematical Physics* (Birkhauser: Bassel).
[11] Suparmi 2011 *Mekanika Kuantum II* (Surakarta: Jurusan Fisika Universitas Sebelas Maret).

Journal of Physics: Conference Series **539** (2014) 012020 doi:10.1088/1742-6596/539/1/012020

Simulation of Drain Currents of Double Gated Armchair Graphene Nanoribbon Field-Effect Transistors by Solving Dirac "Like" Equation and Using Transfer Matrix Method

E Suhendi[1], R Syariati[1], F A Noor[1], N Kurniasih[2], and Khairurrijal[1],#

[1]Physics of Electronic Materials Research Division,
[2]Earth Physics and Physics of Complex Systems Research Division,
Faculty of Mathematics and Natural Sciences, Institut Teknologi Bandung,
Jalan Ganesa 10, Bandung 40132, Indonesia

[#]E-mail: krijal@itb.ac.id

Abstract. We have modelled quantum mechanically the drain current in double gated armchair graphene nanoribbon field-effect transistors (AGNR-FETs). The Dirac 'like' equation was used to determine the electron wave functions in the AGNR. The electron transmittance and the drain current were calculated numerically by using a transfer matrix method (TMM). The results showed that the drain current of the AGNR-FET devices increases with the drain and gate voltages. In addition, the threshold voltage of the devices was obtained to be constant at about 0.3 V. Moreover, the drain current increased with decreasing the thickness of insulator. The different temperatures gave the different characteristics of the drain current.

1. Introduction

Graphene is a sheet of carbon atoms forming two dimensional material systems. Currently, it is still an interesting topic to be theoretically and experimentally studied. Large area graphene has semimetal properties with zero bandgap and high carrier mobility [1]. In addition, the dynamics of electron in graphene for low energy limit is described by the relativistic Dirac equation [2]. When graphene has width smaller than its length, it is called as a graphene nanoribbon (GNR) that has different electrical properties from graphene. Based on the edge shape, GNRs are divided into zigzag GNRs (ZGNRs) and armchair GNRs (AGNRs). The ZGNRs have metal properties for all widths, while the AGNRs can be either semiconducting or metal depending on their widths [3–4]. Because of these properties, GNRs can be applied for various nanoelectronic devices. High speed field-effect transistor (FET) is one of applications of graphene for the nanoelectronic devices [5].

The characteristics of graphene transistors including the tunneling current flowing from the source to the drain have been studied [6-9]. The Schrodinger equation has been used to obtain the characteristics of the devices. In addition, they used the WKB method [6,7] and the non-equilibrium Green's function [8,9] for calculating the drain current. In this paper, we report simulation results of drain current of armchair graphene nanoribbon field-effect transistors (AGNR-FETs) that have been obtained by

[#] Khairurrijal

solving the Dirac 'like' equation and using a transfer matrix method (TMM). The TMM has been employed to compute the tunneling current and it has been shown that the tunneling current calculated under the TMM is better than that done under the WKB [10,11]. The influences of gate and drain voltages, insulator thickness, and temperature on the drain current of (AGNR-FETs) will be discussed.

2. Theoretical Model

The schematic diagram of a double gated AGNR-FET device is shown in Fig. 1. AGNR is placed as a channel, whereas ZGNRs that have metal properties could be used as source and drain of the device.

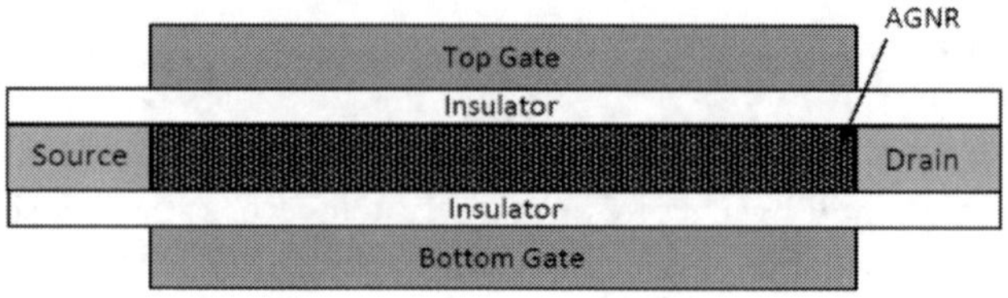

Figure 1. Schematic diagram of a double gated AGNR-FET device.

The electronic structure of AGNRs represented by the dispersion relation, $\varepsilon(\boldsymbol{k})$, is given by [3]

$$\varepsilon = s\hbar v_F \sqrt{k_x^2 + k_n^2}\,, \tag{1}$$

where the transverse momentum, $k_n = n\pi / 3W$, is quantized by the ribbon width in which W is the width of AGNR and $n = \pm1, \pm2, \pm3,\ldots$ $s = +1$ and $s = -1$ indicate the conduction and valence bands, respectively, $\hbar$ is the reduced Planck constant, and v_F=10^6 m/s is the Fermi velocity. The potential profile of the AGNR near the source contact was obtained by solving the Laplace equation from Ref. [6] as given by

$$V(x) = \varphi_{SB} - \frac{2V_{GS}}{\pi}\arccos\left[\exp\left(\frac{-x\pi}{2t_{ins}}\right)\right], \tag{2}$$

where V_{GS} is the gate voltage and t_{ins} is the insulator thickness. The Schottky barrier height, φ_{SB}, was supposed to be half of the AGNR bandgap.
The electron behavior in graphene was described by the Dirac-like Hamiltonian as given by [2]

$$\hat{H} = -i\hbar v_F \sigma.\nabla\,, \tag{3}$$

where σ is the Pauli spin matrix. By solving the Dirac-like Hamiltonian equation, the wave functions in the source, channel and drain region could be found. Furthermore, by employing the TMM in which the potential profile of the AGNR channel was divided into N segments of rectangular form as given in Refs. [10,11] and applying the boundary conditions at each interface, the transmittance through the AGNR channel could be obtained. Finally, by using the Landauer formula which can be written as [6]

$$I_{DS} = \frac{2e}{h}\sum_n \int_{\varphi_{SB}-V_{GS}}^{\varphi_{SB}} T_n(E)\left[f(E) - f(E + eV_{DS})\right]dE\,, \tag{4}$$

where $T_n(E)$ is transmittance of the nth-subband, V_{DS} is the drain voltage, V_{GS} is the gate voltage and $f(E)$ is the Fermi-Dirac distribution function, and by assuming ballistic transport within the AGNR channel, the drain current, I_{DS}, could be determined.

3. Calculated Results and Discussion

The AGNR channel with length of 20 nm and width of 5 nm, the insulator thickness of 2 nm, and the temperature of 300 K were used. The calculated drain currents as a function of the drain voltage for various gate voltages are shown in Fig. 2. It is depicted that initially the drain currents increase with the drain voltage and then they will reach saturation values. It is also noticeable that the typical saturation characteristics are in accordance to the GNR-FETs ones [6-9].

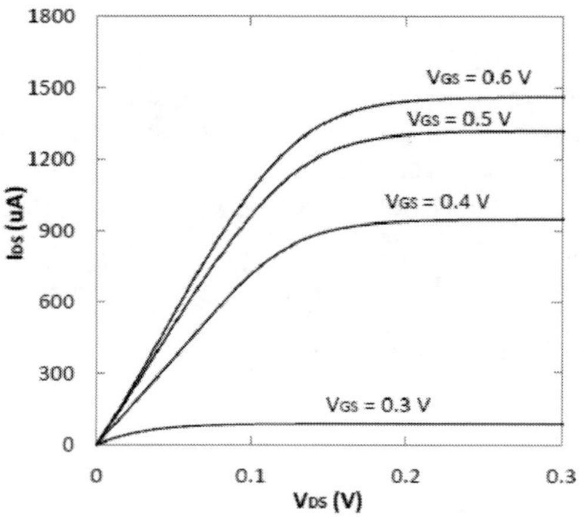

Figure 2. The drain current as a function of the drain voltage for various gate voltages.

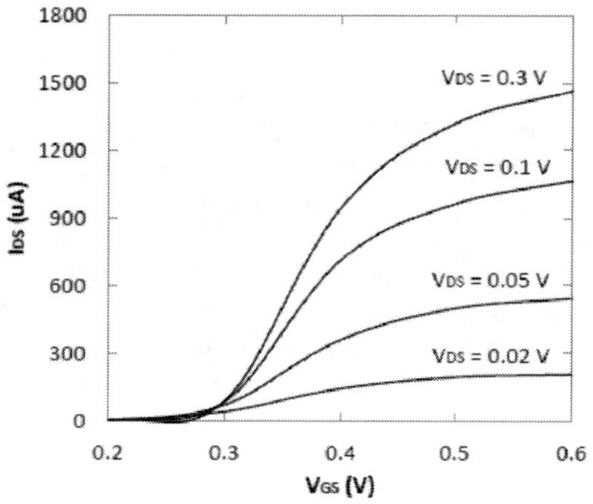

Figure 3. The drain current as a function of the gate voltage for various drain voltages.

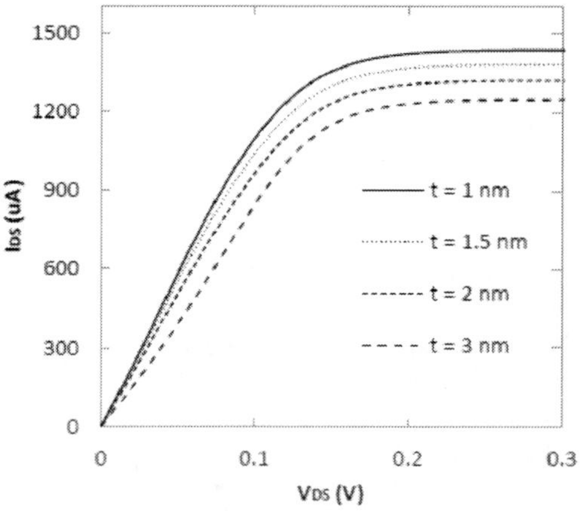

Figure 4. The drain current as a function of the drain voltage for various insulator thicknesses.

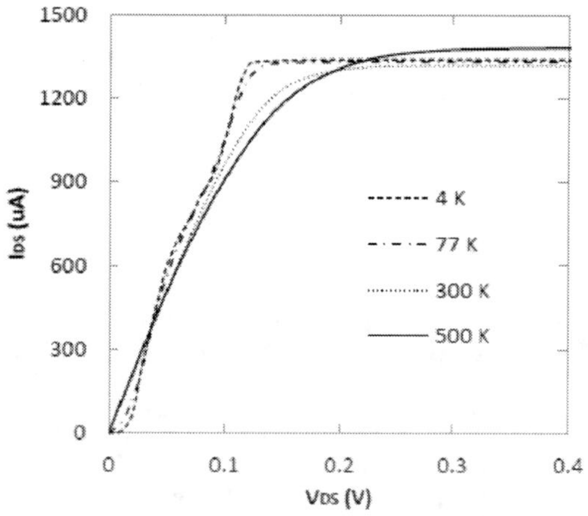

Figure 5. The drain current as a function of the drain voltage for various temperatures.

The dependence of drain current on the gate voltage for different drain voltages is given in Fig. 3. The drain current increases with increasing the gate and drain voltages. Furthermore, it is seen that the threshold voltage is constant about 0.3 V. This finding is in line with that obtained by Ref. [12].

Figure 4 shows the drain current as a function of the drain voltage for different insulator thicknesses. The gate voltage, the AGNR channel length, the AGNR channel width, and the temperature were 0.5 V, 20 nm, 5 nm, and 300 K, respectively. It can be seen that the thinner the insulator thickness the drain current of AGNR-FET becomes higher.

Journal of Physics: Conference Series **539** (2014) 012020 doi:10.1088/1742-6596/539/1/012020

The dependences of drain currents on the drain voltage for various temperatures ranging from 4 K to 500 K are given in Fig. 5. It was taken that the gate voltage, the AGNR channel length, the AGNR channel width, and the insulator thickness were 0.5 V, 20 nm, 5 nm, and 2 nm, respectively. It is shown that the different temperatures give the different characteristics of the drain current of AGNR-FET.

4. Conclusion

The Dirac 'like' equation and the transfer matrix method have been successfully applied to obtain the drain current of AGNR-FETs. It has been found that the drain current increases as the gate and drain voltages increase. The threshold voltage of the devices has been obtained to be constant about 0.3 V. It has also been demonstrated that the drain current increases with decreasing the thickness of insulator. Moreover, the different temperatures give the different characteristics of the drain current of the devices.

5. Acknowledgments

This work was financially supported by "Hibah Desentralisasi", "Riset & Inovasi KK", and "Hibah Kompetensi" Research Grants in the fiscal year 2013-2014.

References

[1] Novoselov K S, Geim A K, Morozov S V, Jiang D, Zhang Y, Dubonos S V, Grigorieva I V and Firsov A A 2004 *Science* **306** 666
[2] Katsnelson M I, Novoselov K S and Geim A K 2006 *Nature Physics* **2** 620
[3] Brey L and Fertig H A 2007 *Phys. Rev.* **B 73** 235411
[4] Son Y W, Cohen M L and Louie S G 2006 *Phys. Rev. Lett.* **97** 216803
[5] Schwierz F 2010 *Nat. Nanotechnol.* **5** 487
[6] Jimenez D 2008 *Nanotechnology* **19** 345204
[7] Kargar A 2009 *9th IEEE Conference on Nanotechnology* 710
[8] Fiori G and Iannaccone G 2007 *Electron Device Letters, IEEE* **28** 760
[9] Ouyang Y, Yoon Y, and Guo J 2007 *Electron Device Letters, IEEE* **54** 2223
[10] Shangguan W Z, Zhou X, Chiah S B, See G H and Chandrasekaran K 2005 *J. Appl. Phys.* **97** 123709
[11] Suhendi E, Syariati R, Noor F A, Kurniasih N and Khairurrijal 2014 submitted to *American Institute of Physics (AIP) Conference Proceedings*
[12] Suhendi E, Noor F A, Kurniasih N and Khairurrijal 2014 *Advanced Materials Research Journal* **896** 367-370

Modeling of Soft sensor based on Artificial Neural Network for Galactic Cosmic Rays Application

W Suparta[1] and W S Putro[1]

[1] Space Science Centre (ANGKASA), Institute of Climate Change, Universiti Kebangsaan Malaysia 43600 Bangi, Selangor Darul Ehsan, Malaysia

E-mail: wayan@ukm.my

Abstract. For successful designing of space radiation Galactic Cosmic Rays (GCRs) model, we develop a soft sensor based on the Artificial Neural Network (ANN) model. At the first step, the soft sensor based ANN was constructed as an alternative to model space radiation environment. The structure of ANN in this model is using Multilayer Perceptron (MLP) and Levenberg Marquardt algorithms with 3 inputs and 2 outputs. In the input variable, we use 12 years data (Corr, Uncorr and Press) of GCR particles obtained from Neutron Monitor of Bartol University (Fort Smith area) and the target output is (Corr and Press) from the same source but for Inuvik area in the Polar Regions. In the validation step, we obtained the Root Mean Square Error (RMSE) value of Corr 3.8670e-004 and Press 1.3414e-004 and Variance Accounted For (VAF) of Corr 99.9839 % and Press 99.9831% during the training section. After all the results obtained, then we applied into a Matlab GUI simulation (soft sensor simulation). This simulation will display the estimation of output value from input (Corr and Press). Testing results showed an error of 0.133% and 0.014% for Corr and Press, respectively.

Keywords: *Galactic Cosmic Rays, Matlab GUI, Artificial Neural Network, Soft sensor*

1. Introduction

One example of space radiation environment source is Galactic Cosmic Rays (GCRs), which is harmful to satellite component. Radiation was composed by different particles (neutron, protons, electron and heavier charged particle) up to 10^{20} eV. Many researchers studied about radiation of GCRs such as Bourdarie and Xapsos [1] showing that GCRs are radiation with high energy and originated from the outside of solar system, which strongly causes Single Event Effect (SEE) in microelectronics and photonics. GCRs dominated the effect in terms of Linear Energy Transfer (LET) in satellite application, and this energy is non-linear variable. Shalchi *et al.* [2] discussed more about nonlinear parallel and perpendicular diffusion of charged cosmic rays whereas his study emphasizes the GCRs is non-linear variable and a new theory for parallel and perpendicular diffusion of cosmic rays. Suparta and Putro [3] discussed an analysis distribution of GCRs in polar orbit satellite for Geant4 (GEometry ANd Tracking) application with simulating the GCRs particle. Chris H *et al.* [4] also discussed a cosmic ray model to generate the input tables used for the Cosmic Ray Shower Generator (CRY) and his finished made a GCR model with this simulation. However, this simulation has found some problem including estimated output model. According to this problem, we found that Artificial Neural Network (ANN) with Levenberg Marquardt algorithm is proposed to estimate and measured the model output.

In this study, we focused to create a soft sensor based on ANN and estimate the ground based Neutron Monitor from Bartol University (Inuvik area) where the input data is taken from the same source (such as Fort Smith area). Matlab software is employed to create ANN and Graphic Utility Interface (GUI) of the soft GCR sensor.

2. Methodology

2.1. Collecting data

The fluxes of GCRs radiation is divided into space (Primary GCRs) and ground based (Secondary GCRs). Figure 1 shows the spectrum of galactic protons flux at Earth, where E (GeV) flux is not correlated with the solar wind. As shown in the lower energy, GCR area is in the left of graph with cut off (~15 GeV at the equator, 0 GeV at the poles) is higher during solar proton minimum and the other side, Earth's magnetic field acts as a further filter on the low-E flux spectrum.

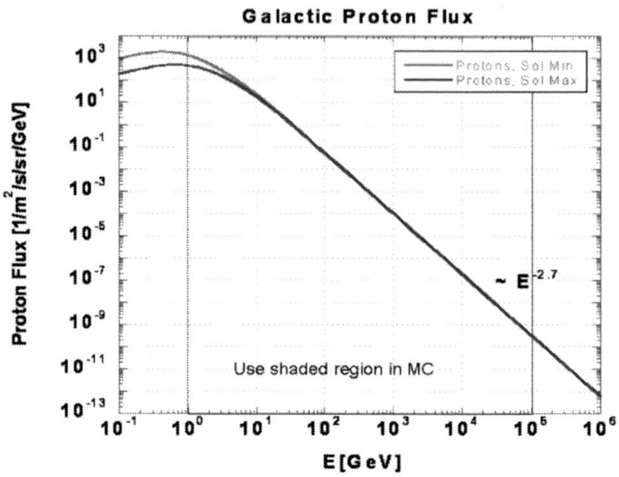

Figure 1. Spectrum of Galactic Protons incident on Earth [6]

In this study, we used 12 years data collected from 1 January 2000 to 31 December 2011. The data was obtained from the ground based neutron monitors at Fort Smith station (60.02°N latitude, 111.93°W and elevation of 206 m) and Inuvik station (Geographic: 68.05°N latitude, 133.70°W and elevation of 21 m) that can be accessed through http://neutronm.bartol.udel.edu/~pyle/bri_table.html. Both ground based Neutron Monitors are located at the North West Territories – Canada. Further, the data was classified of three variables: Corr, Press and Uncorr. Corr data was processed and otherwise is called Uncorr, while Press data is air pressure as shown in Table 1.

Table 1. Ground-base neutron monitors data from Bartol University for two selected stations

Date (Time)	Fort Smith			Inuvik		
yyyy/mm/dd/hh/mm	Corr	Uncorr	Press	Corr	Uncorr	Press
:	:		:	:	:	:
2001/01/01/00/00	6802	8422	738.6	6245	6001	763.1
2001/01/01/01/00	6816	8510	737.8	6238	6000	763.0
2001/01/01/02/00	6838	8543	737.7	6249	6017	762.9
:	:	:	:	:	:	:

2.2. Soft sensor based Artificial Neural Network (ANN)

A soft sensor are used to solve a number of different problems such as measuring system back-up, what-if analysis, real-time prediction for plant control, sensor validation and fault diagnosis strategies Luigi [7]. This sensor can process for many variable measurements simultaneously. Figure 2 shows the identification procedure of a soft sensor design, where in the first step is collecting the data from data base until the last step is a data validation.

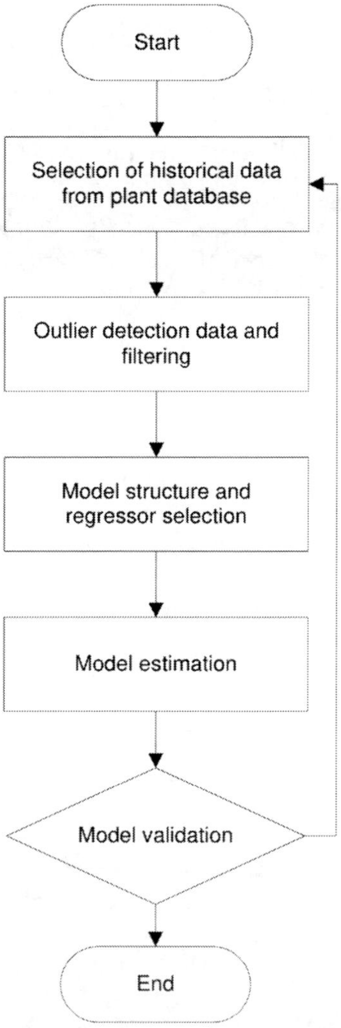

Figure 2. Block scheme of the identification procedure of a soft sensor [7]

There are several methods was used to design soft sensors such as ANN. ANN identification system is an attempt to get a mathematical description (model mathematic) from the dynamical system based on measurement data. It has steps such as experiment, selecting structure model, estimating parameter and model validation. In the experiment section, the variable input and output must be obtained from process description. Below is the common formulas used in the ANN method as described in detail by Laurence [8]. The main idea of the experiment is to combine a varying input u and observe this effect on the output y as shown in Equation 1.

$$Z^N = \{[u(t), y(t)], T = 1,....,N\} \quad (1)$$

After that, select the structure model including structure selection and noise modeling. In this study, we choose Multilayer Perceptron (MLP) for structure selection and ANN with Levenberg Marquardt algorithm for noise modeling. The estimation of parameter a model candidate pairs is expressed in Equation 2.

$$y(t) = y'(t \mid \theta) + e(t) = g[t, \theta] + e(t) \quad (2)$$

The purposes of training are to get a mapping the data pair with candidate models pair as shown by $Z^N \rightarrow \theta$. The data scaling is required to accelerate convergence during training process in ANN method. The average errors from training results between the output process and output targets is calculated using a *Root Mean Square Error* (RMSE). The success of training process was not only examined by RMSE, but in this study we need to compute the value of *Variance Accounted For* (VAF) with the formula given in Equation 3.

$$VAF = \left\{1 - \frac{\mathrm{var}[y(t) - y'(t)]}{\mathrm{var}[y(t)]}\right\} \quad (3)$$

where, *y(t)* and *y'(t)* are experimental input and output variable values.

3. Results and Discussion

Figures 3 and 4 show the input and output variables of GCRs on the ground based Neutron Monitor for 12 years data obtained from Inuvik station (with target output is Corr and Press). The input variable in this study is Corr, Press and Uncorr) from the Fort Smith station.

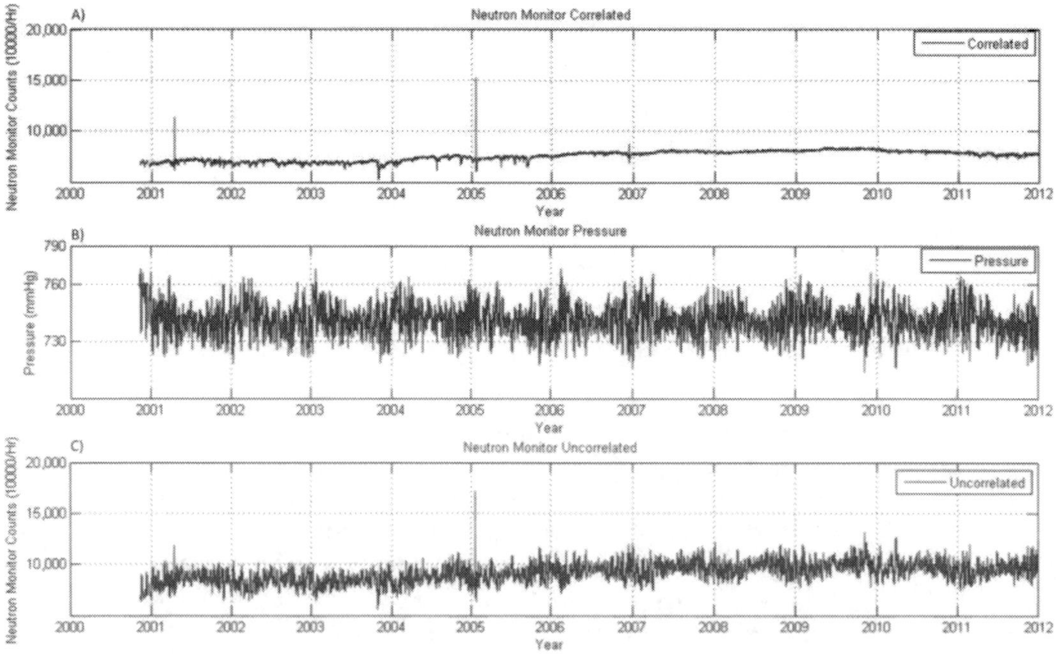

Figure 3. Data input for modelling soft GCR sensor from Fort Smith area for data (a) Corr, (b) Press and (c) Uncorr

As shown in the figure, data input on January 2000 was missing because neutron monitor in the Fort Smith area was under construction. Therefore, in the training process a missing data (or NaN) should be deleted. The neutron monitors in this area have been developed in October 2000. However, on July 2000, the neutron monitor at Inuvik station was detected too much neutron as shown in Figure 4 (between 2000 and 2002). Because *"The Bastille Day Event"* occurred in July 2000 and *"The Halloween solar storms"* in October 2003, the prominent peak of Corr data was detected a lot of neutron due these solar phenomena, both Fort Smith and Inuvik areas (January 2005).

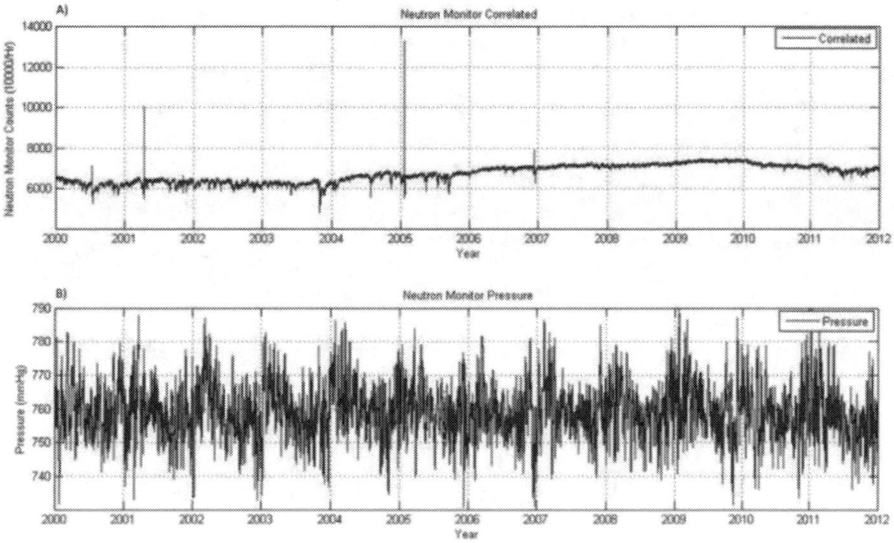

Figure 4. Data output for model at Inuvik area for (a) Corr and (b) Press

In this study, we would to create ground based data based on the ANN model and develop Graphic Utility Interface (GUI) for a soft sensor as shown in Figures 5 and 6. In the training process, we use MLP structure, which have three layers: input, hidden and output.

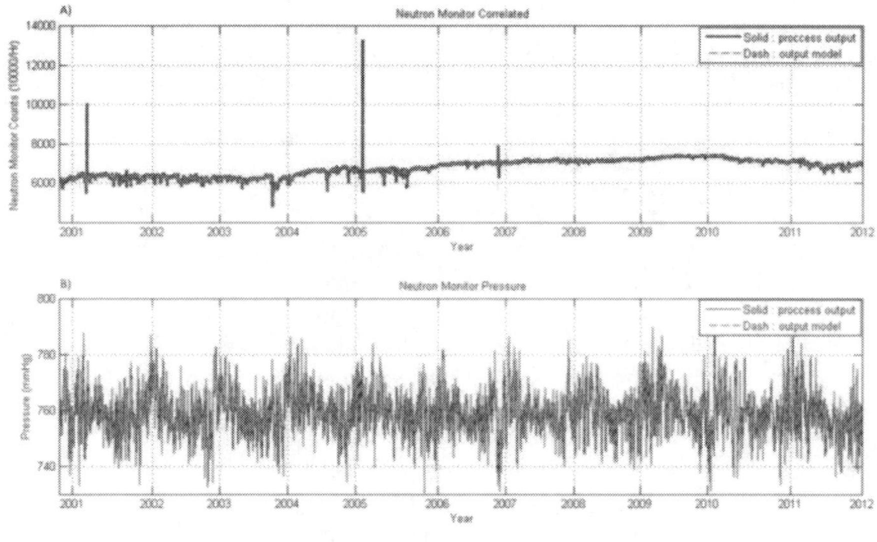

Figure 5. Result of training model based ANN with Levenberg Marquardt algorithm for (a) Corr and (b) Press

As shown in the figure, the red line is result of training and the blue lines are target (output variable). In the result of training process (red line), these line is always tracking on blue line color, this indicate that the training process has been successful and the model has a tendency to follow the pattern of data obtained from target.

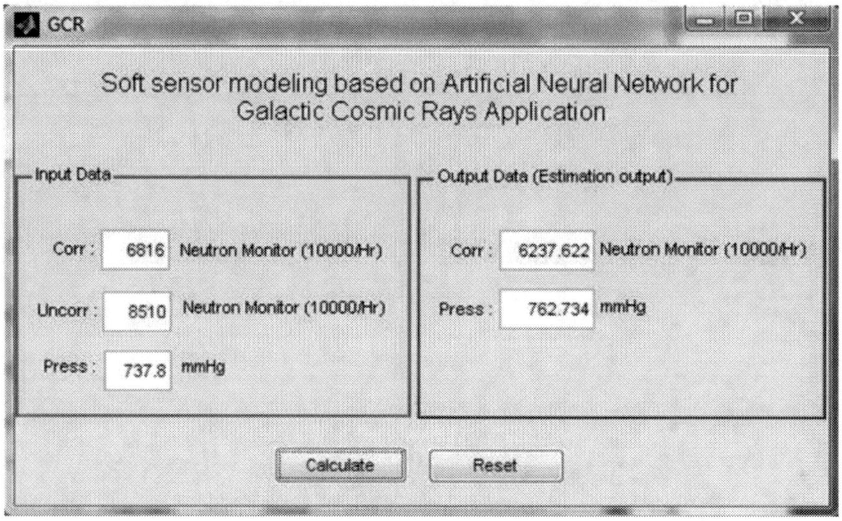

Figure 6. The soft sensors GUI

As shown in the figure, the first step is to develop a soft sensors GUI are by inserting in command window Matlab software `guide` and choose GUI with Uicontrols. The second steps, put all functions at training section to calculate button and the last step create some functions to clear edit text on red and blue box with reset button. The red box is input data and blue box is result of estimation from input variable. For example we choose corr 6816 Neutron Monitor (10^4/Hour), uncorr 8510 Neutron Monitor (10^4/Hour) and press 737.8 mmHg (see Table 1, Column 3). The result of estimation input variable is corr 6237.622 Neutron Monitor (10^4/Hour) and press 762.734 mmHg. We use *three* hidden layer and *linear* activation function on construction of hidden layer. The reason is based on the value of iteration, *Root Mean Square Error* (RMSE) and *Variance Accounted For* (VAF) in the training process as shown Table 3 and Table 4.

Table 3. Iteration result

Iteration	Weight
1.	7.216e-003
2.	3.681e-005
3.	4.388e-006
4.	4.236e-006
5.	4.199e-006
6.	4.185e-006
7.	4.176e-006
8.	4.173e-006
9.	4.165e-006
10.	4.164e-006
11.	4.164e-006
12.	4.164e-006
13.	4.164e-006
14.	4.164e-006

Table 4. RMSE, VAF and Percent error result

Output Model	Result		
	RMSE	VAF (%)	Percent error (%)
Corr	3.8670e-004	99.9839	0.133
Press	1.3414e-004	99.9831	0.014

4. Conclusion

We had successful developed a soft sensor based on the Artificial Neural Network (ANN) model from ground based Neutron Monitor data. The soft sensor to simulate of Galactic Cosmic Rays (GCRs) using the 12 years data and is performed with Matlab Graphic Utility Interface (GUI) software. In the training process, we use Multilayer Perceptron (MLP) structure that to have three layers. The result on training process from ANN model is examined using *Root Mean Square Error* (RMSE) and *Variance Accounted For* (VAF). The maximum percent error values are 0.133% and 0.014% for Corr and Press, respectively. From this results showed that the soft sensor developed showed with high accuracy and capable to be used for GCRs estimation. The next study we will to do a modeling of GCRs radiation for a satellite application in the equatorial region like Near Equatorial Orbit (NEqO) to assess the GCRs impact on our spacecraft.

Acknowledgments

This research is supported by Ministry of Education Malaysia (MoE) under grant FRGS/2/2013/SG03/UKM/02/3. The authors would like thank Bartol University for providing data used in this study, and Matlab forums for advising in Matlab GUI programming under Windows platform.

References

[1] Bourdarie S and Xapsos X 2008 *IEEE Transactions on Nuclear Science* **55** 1810
[2] Shalchi A 2004 *The Astrophysical Journal* **616** 617
[3] Suparta W and Putro W S 2014 *Procedding 2014 International Conference On Science & Enggineering In Mathematics, Chemistry and Physics* 323
[4] Chris H 2007 *IEEE Nuclear Science Symposium Conference Record* 1143
[5] Trauth M H 2010 MATLAB® Recipes for Earth Sciences, third edition Springer – Verlag New York p 235
[6] Papini P 1996 *Nuovo Cimento C Geophysics Space Physics C* 367
[7] Luigi F 2006 *Soft Sensors for Monitoring and Control of Industrial Processes* (New York: Springer - Verlag)
[8] Laurence F 1994 Fundamental of Neural Network Prentice Hall Inc. 79

Journal of Physics: Conference Series **539** (2014) 012022 doi:10.1088/1742-6596/539/1/012022

Shock wave propagation of circular dam break problems

Sudi Mungkasi

Department of Mathematics, Faculty of Science and Technology, Sanata Dharma University, Mrican, Tromol Pos 29, Yogyakarta 55002, Indonesia

E-mail: sudi@usd.ac.id

Abstract. We examine the behavior of shock wave propagation of circular (radial) dam break problems. A dam break problem represents a reservoir having two sides of water at rest initially with different depth separated by a wall, then water flows after the wall is removed. The behavior of shock wave propagation is investigated with respect to water levels and with respect to the speeds of the shock waves. To the author's knowledge, such investigation for circular dam break problems had never been done before. Therefore, this new work shall be important for applied computational mathematics and physics communities as well as fluid dynamic researchers. Based on our research results, the propagation speed of shock wave in a circular dam break is lower than that of shock wave in a planar dam break having the same initial water levels as in the circular dam break.

1. Introduction

Water can flow in either a closed or open space. An example of water motion in a closed channel is pipe flows. An example of water motion in open space is flood flows. Studies of water flows are important, as they occur in many situations or conditions (see References [1]-[6]).

This paper considers water flows in an open channel. We solve a circular dam break problem. It is also known as a radial dam break problem. A circular dam represents a water reservoir having a wall with circle in shape, where the depth of water inside the circle wall is greater than that of water outside. Then the circular dam break problem means that we need to find the properties (water surface, momentum, velocity, energy, etc.) of water after the circular dam wall is removed completely at an instant of time. We assume that initially the area outside of the circular wall has a positive constant depth. Therefore when dam break happens, a shock wave appears and propagates radially [3].

The circular dam break problem can be modelled by the the one-dimensional shallow water equations with varying width as well as the standard two-dimensional shallow water equations. A simulation of the problem through the one-dimensional shallow water equations with varying width was conducted by Roberts and Wilson [5]. Some simulation results of the problem through the standard two-dimensional shallow water equations was presented by Mungkasi [4]. Shallow water flows was modelled mathematically for initial stages by Saint-Venant [2].

Our goal is to research on the shock wave propagation of the circular dam break problem. We implement the one-dimensional shallow water equations with varying width following Roberts and Wilson [5]. An advantage of using these equations is that the numerical method (used

Journal of Physics: Conference Series **539** (2014) 012022 doi:10.1088/1742-6596/539/1/012022

to solve these equations) is simpler than the two dimensional version. It is because the one-dimensional shallow water equations with varying width have the same form as the standard one-dimensional shallow water equations.

The rest of this paper is organized as follows. Governing equations are recalled in Section 2. The numerical method of Roberts and Wilson [5] is briefly presented in Section 3. Then Section 4 presents and discusses numerical results on the shock wave propagation. Finally we conclude our presentation with some remarks in Section 5.

2. Governing equations

The one-dimensional shallow water equations with varying width are [5]

$$\frac{\partial}{\partial t}(bh) + \frac{\partial}{\partial x}(bhu) = 0\,, \tag{1}$$

$$\frac{\partial}{\partial t}(bhu) + \frac{\partial}{\partial x}(bhu^2 + \frac{1}{2}gh^2 b) = -gh\frac{dz}{dx}b + \frac{1}{2}gh^2\frac{db}{dx}\,. \tag{2}$$

Here $h(x,t)$ is water depth, $u(x,t)$ is horizontal velocity, $z(x)$ is the topography, $b(x)$ is the channel varying width and g is the accelleration due to gravity. The free variables are time t and space x. The absolute water level is called stage and defined extensively as $w = h + z$.

Some notes are as follows. When the width $b(x)$ is constant, then the shallow water equations (1) and (2) are simplified to the standard one-dimensional shallow water equations. When we have horizontal topography, the source term $-gh\frac{dz}{dx}b$ disappears as $\frac{dz}{dx} = 0$.

3. Numerical method

The finite volume method of Roberts and Wilson [5] is used to solve the shallow water equations (1) and (2). The method is briefly described as follows.

Consider equations (1) and (2). These equations are conservation laws of the form

$$\frac{\partial}{\partial t}\mathbf{q} + \frac{\partial}{\partial x}\mathbf{f}(\mathbf{q}) = \mathbf{s}\,. \tag{3}$$

Here $\mathbf{q}$ is the vector of conserved quantities, $\mathbf{f}(\mathbf{q})$ is the vector of fluxes and $\mathbf{s}$ is the vector of sources.

Assume that we are given a space domain. A discretization of the space domain leads equation (3) to the semi-discrete finite volume scheme

$$\frac{d}{dt}\mathbf{q}_i + \frac{1}{\Delta x_i}\left(\mathbf{f}_{i+\frac{1}{2}} - \mathbf{f}_{i-\frac{1}{2}}\right) = \mathbf{s}_i\,. \tag{4}$$

The accuracy of the finite volume method is then dependent on the accuracy of the numerical fluxes $\mathbf{f}_{i+\frac{1}{2}}$ and $\mathbf{f}_{i-\frac{1}{2}}$ as well as on the accuracy of the solver of the ordinary differential equation (4). For our simulations we use second order finite volume method. It is second order accurate in space and second order accurate in time. We implement the minmod limiter to overcome artificial oscillation of numerical solutions.

For more details on this finite volume method for solving the shallow water equations (1) and (2), we refer to Roberts and Wilson [5].

4. Numerical results

To achieve the goal of this paper we consider a circular dam break problem. All quantities are given in SI units, so we omit the writing of units as they are already clear.

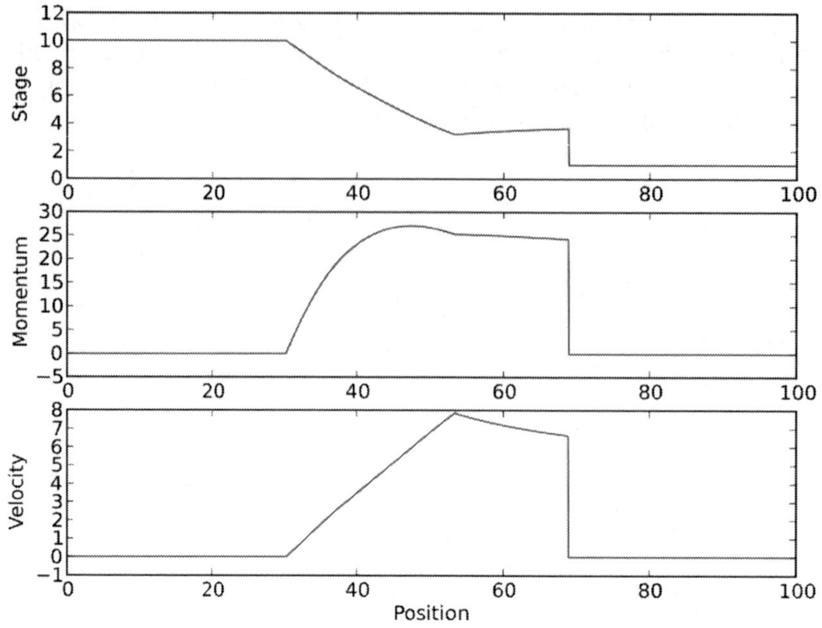

Figure 1. Simulation results for time $t = 2$.

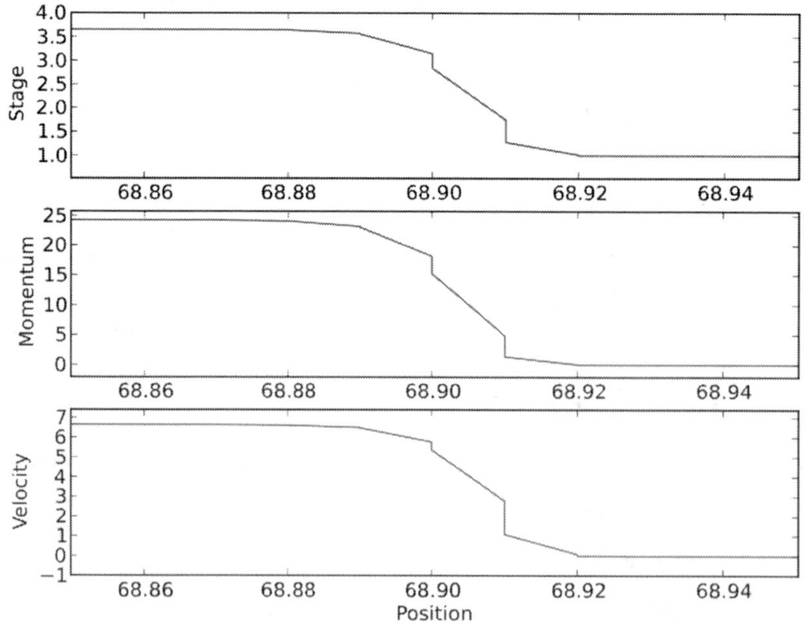

Figure 2. A magnification around shock position for time $t = 2$.

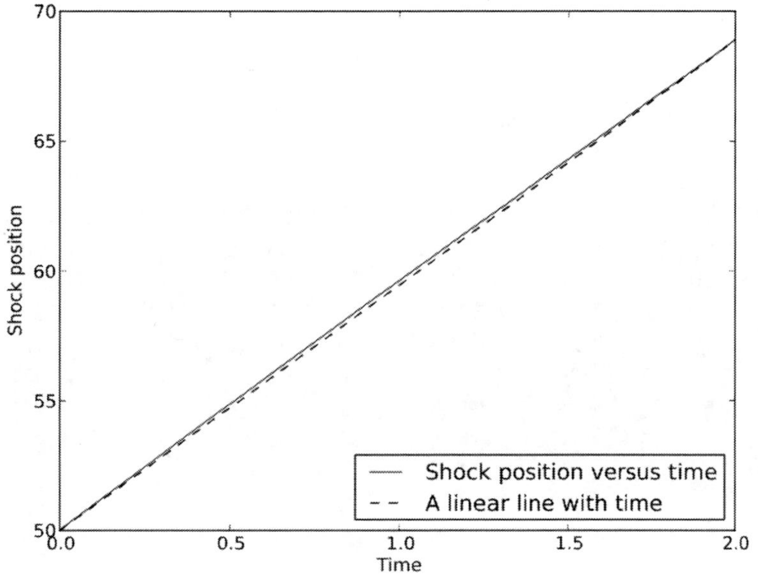

Figure 3. A track of the shock front with respect to time for $t \in [0, 2]$. The solid line shows the shock position versus time. The dashed line shows a linear line connecting the initial point $(0, 50)$ and the final point $(2, 68.915)$ of the shock positions.

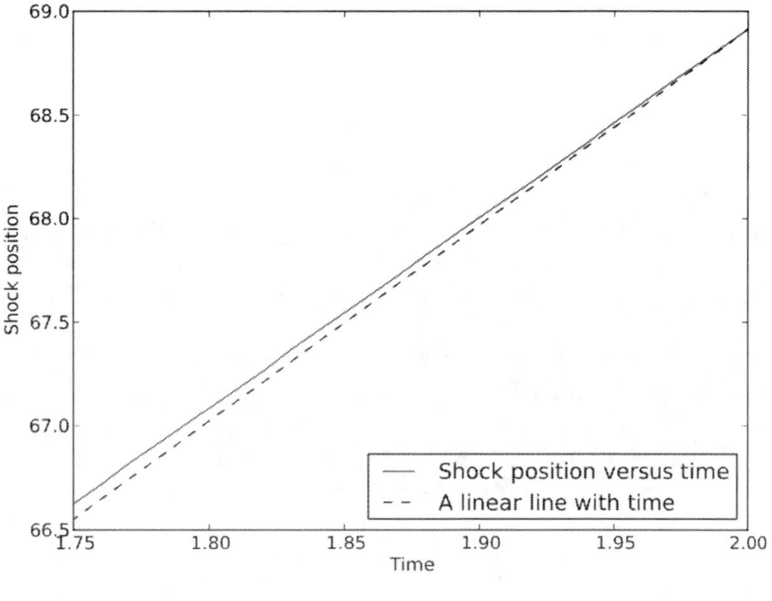

Figure 4. A track of the shock front with respect to time for $t \in [1.75, 2]$. The solid line shows the shock position versus time. The dashed line shows a linear line connecting the initial point $(0, 50)$ and the final point $(2, 68.915)$ of the shock positions.

Journal of Physics: Conference Series **539** (2014) 012022 doi:10.1088/1742-6596/539/1/012022

Table 1. Track of the shock front with respect to time for $t \in [0, 1]$.

time	position	time	position	time	position	time	position
0.01	50.115	0.26	52.555	0.51	54.975	0.76	57.365
0.02	50.215	0.27	52.655	0.52	55.065	0.77	57.455
0.03	50.315	0.28	52.745	0.53	55.165	0.78	57.555
0.04	50.415	0.29	52.845	0.54	55.255	0.79	57.645
0.05	50.505	0.3	52.945	0.55	55.355	0.8	57.745
0.06	50.605	0.31	53.045	0.56	55.455	0.81	57.835
0.07	50.705	0.32	53.135	0.57	55.545	0.82	57.935
0.08	50.805	0.33	53.235	0.58	55.645	0.83	58.025
0.09	50.895	0.34	53.335	0.59	55.735	0.84	58.125
0.1	50.995	0.35	53.425	0.6	55.835	0.85	58.215
0.11	51.095	0.36	53.525	0.61	55.925	0.86	58.315
0.12	51.195	0.37	53.625	0.62	56.025	0.87	58.405
0.13	51.295	0.38	53.715	0.63	56.125	0.88	58.505
0.14	51.385	0.39	53.815	0.64	56.215	0.89	58.595
0.15	51.485	0.4	53.915	0.65	56.315	0.9	58.695
0.16	51.585	0.41	54.005	0.66	56.405	0.91	58.785
0.17	51.685	0.42	54.105	0.67	56.505	0.92	58.885
0.18	51.775	0.43	54.205	0.68	56.595	0.93	58.975
0.19	51.875	0.44	54.295	0.69	56.695	0.94	59.075
0.2	51.975	0.45	54.395	0.7	56.795	0.95	59.165
0.21	52.065	0.46	54.485	0.71	56.885	0.96	59.255
0.22	52.165	0.47	54.585	0.72	56.985	0.97	59.355
0.23	52.265	0.48	54.685	0.73	57.075	0.98	59.445
0.24	52.365	0.49	54.775	0.74	57.175	0.99	59.545
0.25	52.455	0.5	54.875	0.75	57.265	1	59.635

Similar to Birman and Falcovitz [1], we consider the shallow water equations (1) and (2) with varying width $b(x) = 2\pi x$. The length of the channel is 100. Stage is 10 for $x \in [0, 50]$. However stage is 1 for $x \in [50, 100]$. (Roberts and Wilson [5] set stage to be 2 for $x \in [50, 100]$.) This problem mimics the two-dimensional circular dam break problem. We can solve this problem using equations (1) and (2), as we exploit the symmetry of the water motion after dam break. The acceleration due to gravity is set to 9.81. The space domain $[0, 100]$ is discretized into 1000 cells uniformly.

Figure 1 shows the simulation results at time $t = 2$. The first subfigure is the stage. The second and third subfigures are momentum and velocity respectively. We can see that there is no constant velocity across the moving water. To see the shock front clearer, we magnify Figure 1 around the shock position. The magnification is shown in Figure 2.

Moreover, the track of the shock front is summarized in Table 1 for time $t \in [0, 1]$ and Table 2 for time $t \in [1, 2]$. To see the relationship between the shock track with respect to time, we plot the data of Tables 1 and 2 into Figure 3. A magnification of Figure 3 for time $t \in [1.75, 2]$ is given in Figure 4. From Figures 3 and 4 we conclude that the relationship between the shock wave propagation with respect to time is nonlinear. This means that the shock speed of the circular dam break problem is not constant.

This is different phenomena from a corresponding planar dam break problem. Recall that for the planar dam break problem, the relationship between the shock wave propagation with respect to time is linear. That is due to the constant shock wave propagation.

We note that after 2 seconds of the circular dam break, the shock front travels about 18.915 m.

Table 2. Track of the shock front with respect to time for $t \in [1,2]$.

time	position	time	position	time	position	time	position
1.01	59.735	1.26	62.075	1.51	64.405	1.76	66.715
1.02	59.825	1.27	62.175	1.52	64.505	1.77	66.815
1.03	59.925	1.28	62.265	1.53	64.595	1.78	66.905
1.04	60.015	1.29	62.365	1.54	64.685	1.79	66.995
1.05	60.105	1.3	62.455	1.55	64.775	1.8	67.085
1.06	60.205	1.31	62.545	1.56	64.875	1.81	67.175
1.07	60.295	1.32	62.645	1.57	64.965	1.82	67.265
1.08	60.395	1.33	62.735	1.58	65.055	1.83	67.365
1.09	60.485	1.34	62.825	1.59	65.145	1.84	67.455
1.1	60.575	1.35	62.925	1.6	65.245	1.85	67.545
1.11	60.675	1.36	63.015	1.61	65.335	1.86	67.635
1.12	60.765	1.37	63.105	1.62	65.425	1.87	67.725
1.13	60.865	1.38	63.205	1.63	65.515	1.88	67.825
1.14	60.955	1.39	63.295	1.64	65.615	1.89	67.915
1.15	61.045	1.4	63.385	1.65	65.705	1.9	68.005
1.16	61.145	1.41	63.475	1.66	65.795	1.91	68.095
1.17	61.235	1.42	63.575	1.67	65.885	1.92	68.185
1.18	61.335	1.43	63.665	1.68	65.985	1.93	68.275
1.19	61.425	1.44	63.755	1.69	66.075	1.94	68.365
1.2	61.515	1.45	63.855	1.7	66.165	1.95	68.465
1.21	61.615	1.46	63.945	1.71	66.255	1.96	68.555
1.22	61.705	1.47	64.035	1.72	66.345	1.97	68.645
1.23	61.795	1.48	64.135	1.73	66.445	1.98	68.735
1.24	61.895	1.49	64.225	1.74	66.535	1.99	68.825
1.25	61.985	1.5	64.315	1.75	66.625	2	68.915

However if we set a corresponding planar dam break problem (having the same initial water levels as in the circular dam break), based on the analytical solution of Stoker [6], the shock speed of the corresponding planar dam break is 9.8193. Therefore after 2 seconds of the planar dam break, the shock travels a distance of 19.6386. That is, the propagation speed of shock wave in a circular dam break is lower than that of shock wave in a planar dam break having the same initial water levels as in the circular dam break. Note that a corresponding planar dam break for our simulation means that we have a space $x \in [-100, 100]$, where the initial still water depth is 10 for $x \in [-100, 0]$ and 1 for $x \in [0, 100]$.

5. Conclusions

We have presented research results on the shock wave propagation of a circular dam break problem. We find that the relationship between the shock front of the circular dam break problem with respect to time is nonlinear. This nonlinear phenomenon is due to the nonconstant shock speed of the circular dam break problem. In addition, we find that a corresponding planar dam break problem having the same initial water levels as in the circular dam break produces a faster shock propagation.

Acknowledgments

The author thanks Associate Professor Stephen Gwyn Roberts and Mr Padarn Wilson of The Australian National University for discussion on the shallow water equations with varying width.

Journal of Physics: Conference Series **539** (2014) 012022 doi:10.1088/1742-6596/539/1/012022

References

[1] Birman A and Falcovitz J 2007 *J. Comput. Phys.* **222** 131
[2] de Saint-Venant B 1871 *C. R. Acad. Sci. Paris* **73** 147
[3] LeVeque R J 2002 *Finite Volume Methods for Hyperbolic Problems* (Cambridge: Cambridge University Press)
[4] Mungkasi S 2013 *Bull. Aust. Math. Soc.* **88** 351
[5] Roberts S G and Wilson P 2011 *ANZIAM J.* **52** C967
[6] Stoker J J 1957 *Water Waves: The Mathematical Theory with Applications* (New York: Interscience)

The Variability of Space Radiation Hazards towards LEO Spacecraft

Wayan Suparta[1], **Gusrizal**[1]

[1]Space Science Centre (ANGKASA), Institute of Climate Change, Universiti Kebangsaan Malaysia, 43600 Bangi, Selangor Darul Ehsan, Malaysia

E-mail: wayan@ukm.edu.my

Abstract. The variability of space radiation hazard towards the low-Earth orbit (LEO) spacecraft is reviewed. Three major space radiations, i.e. galactic cosmic rays (GCRs), solar proton events (SPEs), and trapped particles form potential hazards for LEO spacecraft. We focused on trapped particle hazard due to its major role for LEO spacecraft anomaly. We examined the trapped particle variability distribution in solar cycle 23. The year of 2000 and 2008 were chosen to best represent solar maximum and minimum in that cycle. Monthly average particles fluxes plotted in both years showed a quite similar value for all peak values during solar maximum. We also examined events of LEO satellites anomaly for both years by using space weather and geomagnetic parameters as well as the distribution of particles fluxes during those events. Analysis showed that all parameter indicators in the year 2000 were in a critical condition, whereas for the year of 2008, the anomaly occurs was in a normal state.

1. Introduction

There are three major space radiation components in the Sun-Earth environment, i.e. Galactic Cosmic Rays (GCRs), Solar Proton Events (SEPs) and trapped particles [1]. Those three radiations form potential hazards for spacecraft operation as well as a space crew [2]. GCRs are produced in the supernova remnant by diffusive shock acceleration and distributed uniformly in the whole galaxy [3]. The energy of GCRs is in the range of 10^1 to 10^{12} MeV in the interstellar space and rises up to 1 GeV in our solar system [2]. GCRs are divided into primary and secondary cosmic ray. Primary GCRs is a result of the astrophysical process, largely consist of protons, alpha particles (99%), and a small number of heavy nuclei. Secondary GCRs is the result of the decay of primary cosmic ray when it interacts with the Earth's atmosphere. Secondary GCRs is made up of neutrons, pions, positrons, and muons. The GCRs with energy below 1 GeV are influenced by the solar cycle and are attenuated while interacting with the solar wind. The flux of GCRs reaches a minimum during the solar maximum and is maximized during a solar minimum. This occurrence is related to the intensity of solar wind. Since GCRs consist of charged particles, they are also affected by the Earth magnetic field. According to the Lorentz force law, the particle that is parallel to the magnetic field line follows the line and the particle that moves perpendicular to the line will be deflected. Hence, the GCRs are funneled toward the pole and deflected at the Equator. Related to this mechanism, an LEO satellite receives great exposure of GCRs while it flies by near the pole and face minimum GCRs exposure while in the Equator.

The Sun flare and CMEs produce electrons, protons, and heavier charged particles during their lasts. Due to its abundance and high energy, proton dominates more than the other particles. Proton that is erupted during flares, and CMEs is called a solar proton event (SPE) when their energies > 30 MeV of 10^6 up to 10^7 cm^{-2}, whereas a major SPE when their energies > 10 MeV of 10^{10} cm^{-2}. SPEs occurrences are related to solar variability and usually happened during solar maximum. The frequency of ordinary SPE during a solar cycle is ~ 50 events whereas the major SPE occurs is about 1-2 events in a cycle. There are some different characteristics of SPEs produced by flares and CMEs. SPEs produced by flares usually have short life time and release large fluxes of electron. The flare's SPE also has a small fluence (10^7 to 10^8 cm^{-2}) with restricted solar longitude angle (30-45°). On the other hand, the CME's SPE has a longer life time, high fluence (10^9 cm^{-2}), and wider solar longitude angle (from 60° to 180°). Figure 1 shows the conceptual mechanism of Earth's space radiations.

As shown in the right side of Fig. 1, trapped particles are produced by a constant solar plasma stream that flows to the Earth atmosphere and which is carried by the solar wind. These particles are trapped by the Earth's atmosphere and being deposited into two layers, the inner and the outer belts, which are well known as the Van Allen radiation belts. Trapped electron presents the inner belt as well as at the outer layer. The inner belt electron populates in the area that extends to about 2.4 Earth radii and mostly dominated by electron that has less than ~5 MeV of energy. An LEO satellite has to interact with the inner belt particles especially at the South Atlantic Anomaly (SAA) region. The SAA is a region of decreased magnetic field over southern Atlantic Ocean and South America [4]. In addition, the outer electron range is about 2.8 to 12 Earth radii with energy up to ~7 MeV. On the other hand, trapped proton only presents in the inner layer of Van Allen belt with energy ranging from several keV to several hundred of MeV. The variability of both trapped particles is influenced by the solar activity. The inner belt proton is higher during a solar minimum whereas the outer belt electron is higher during a declining phase of solar activity [5].

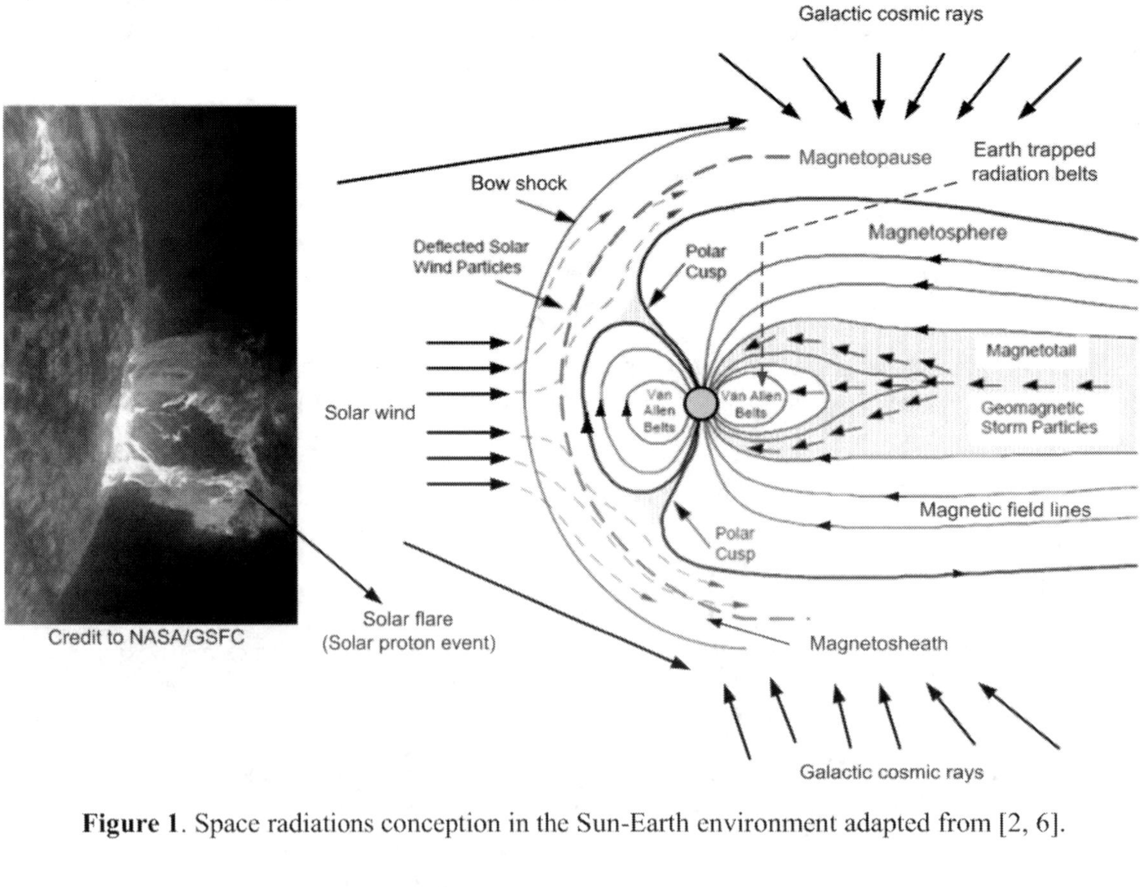

Figure 1. Space radiations conception in the Sun-Earth environment adapted from [2, 6].

Journal of Physics: Conference Series **539** (2014) 012023 doi:10.1088/1742-6596/539/1/012023

The potential hazards of GCRs for the LEO spacecraft are the phenomena of single event effects (SEEs) [7]. The most dangerous component of GCRs is the abundant deposit of Fe nuclei. This heavy ion could trigger the SEEs in space electronic systems. SEEs are occurred when a charged particle is collected on spacecraft circuits such as memories, power devices, logic devices, and disrupting their operation. There are different types of SEEs such as the single even upsets (SEU), single event transients (SET), single event functional interrupts (SEFI), and multiple bit upsets (MBU). The SEU, a SEEs variant that occurred in memory, is the most known and recognized hazard from GCRs to the spacecraft. Not only trapped particles harmful to spacecraft charging, SPEs have a combination effect of GCRs and trapped particles for an LEO spacecraft. It could cause SEEs and charging effect because it triggers the particle precipitation to the Van Allen belt during solar storm. Table 1 summarizes the space radiation effects on LEO satellites [2, 8].

Table 1. The summary of space radiation hazards on LEO satellites

Radiation type	**Effects on space craft**	**Domination in LEO**
Galactic Cosmic Rays (GCRs)	Upsets in electronics, interference with sensor.	Dominate at high inclination orbit (Polar).
Solar Proton Events (SPEs)	Radiation damage in various kind, upsets in electronics, and massive interference with sensor.	More severe for high inclination orbit.
Trapped particles	Surface and interior charging, radiation damage in various kind, upsets in electronics, and considerable interference with sensor	Dominate at low inclination orbit (Equator) with high altitude.

2. Trapped Particles Distribution Variability

Considering dominant factors of trapped particles for LEO satellite anomaly summarized in Table 1, we focus our analysis on trapped particle distribution on solar activity variability. We choose solar cycle 23 as our base for analysis, and we take the year of 2000 and 2008 as the representative of solar maximum and minimum, respectively. Figure 2 presents flux variability of several electrons and protons in different energy levels, taken by National Oceanic and Atmospheric Administration (NOAA) in years mentioned. The energy levels of electron are mep0e1 (> 30 keV), mep0e2 (> 100 keV), and mep0e3 (> 300 keV). In addition, the energy levels for proton are mep0p1 (30-80 keV), mep0p2 (80-240 keV), mep0p3 (240-800 keV), mep0p4 (800-2500 keV), mep0p5 (2500-6900 keV), and mep0p6 (> 6900 keV). The NOAA 15 satellite has an altitude of 807 km with an inclination angle of 98.5°. Therefore, this satellite is measuring the flux distribution in the inner radiation belt. Details of NOAA 15 and the sensor used to measure those particles flux can be found in [9].

Figures 2a and 2b show that the fluxes of lower energy level such as mep0e1, mep0e2, and mep0p1 have higher flux values than other energy levels. It is also seen that low electron fluxes dominate the inner radiation belt, although the layer is a habitual area for trapped proton. The maximum fluxes of all particles are also quite a similar for both years. This is because the particle tends to enter the stable condition once they enter the inner layer, compared to the outer layer that is influenced directly by the sun and astrophysical processes [10].

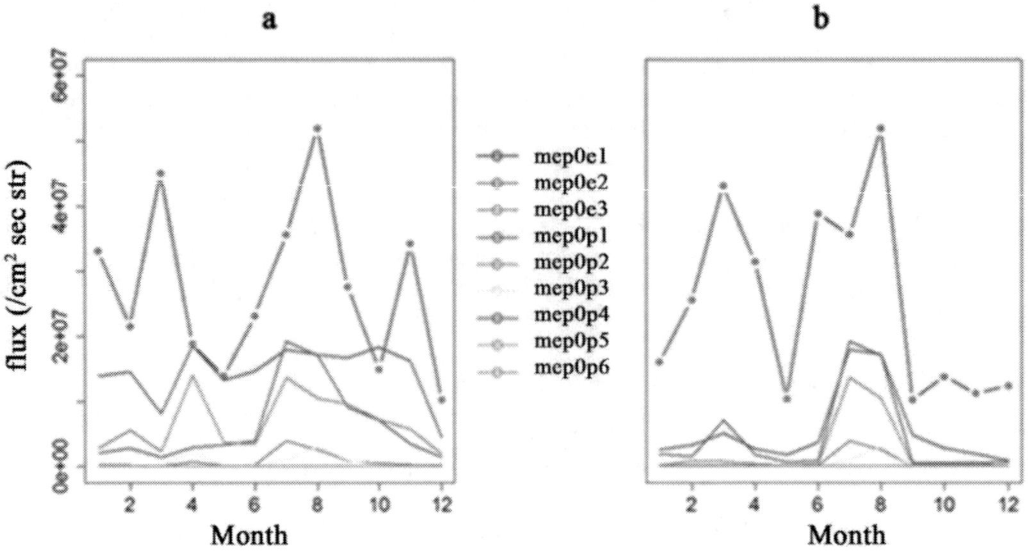

Figure 2. Trapped particles monthly average plots for the year of (a) 2000 and (b) 2008.

3. Trapped Particles and LEO Satellites anomaly

In order to examine the influence of trapped particles towards a satellite anomaly, we study the event of two satellites anomaly occurred in solar maximum and minimum phases. Two LEO satellite anomalies are chosen for this study, i.e. Astro-D and Orbcomm-1. The Astro-D, a LEO satellite with 140 km of altitude and 31° of inclination experienced an anomaly on 15 July 2000 whereas the Orbcomm-1 (altitude of 758 km and 45° of inclination) faced an anomaly on 10 November 2008 (http://sat-nd.com/failures/). We also study several space weather parameters such as solar radio wave (F10.7), solar wind flow speed (FS), and geomagnetic indexes (Ap, Dst and AE) as shown in Figs. 3a-b and 4a-b. All these parameters data can be found in http://omniweb.gsfc.nasa.gov. The particles flux distributions are also examined for those satellites anomaly event. Distribution maps from http://ngdc.noaa.gov/sem/poes/data are chosen for this purpose (as in Figs. 3c-d and 4c-d). We select date of 13-16 July 2000 for Astro-D failure and 8-11 November 2008 for the Orbcomm-1. In this study, only mep0e1 and mep0p2 distributions were selected for simplifying purpose. By this stride, we aim to study the variability distribution of trapped particles before and after the anomaly events.

From Fig. 3c-d, we find that increased radiation in the upper atmosphere occurred a few days before and after Astro-D satellite anomaly was reported. A F10.7 index reached a maximum value of about 325 sfu and accompanied by Joule heating in which the AE index reached a maximum of about 2,023 nT. Solar wind speed fluctuated during this period and reached a maximum of 1,107 km/s that allegedly caused the injection of plasma in the Earth's magnetosphere. Increased magnetic activity which triggered a large-scale extreme storm can be seen in the Dst index with minimum value of -301 nT. The magnetic storm caused changes in low orbit where the plasma electron and proton fluxes varies from 10^3 to 10^6 particles/cm^2 sec str. Magnetic storms also cause dynamic flux distribution for both electron and proton. It can be seen that along the equatorial region the particle precipitation occurred on a large scale with widespread pattern and it concentrated in the SAA area.

The date of 8-11 November 2008 was a period of the lowest solar activity in solar cycle 23 since the minimum phase in this cycle was occurred around this date. From Fig. 4, we find that solar wind speed only reaches a maximum of about 418 km/s. A F10.7 radiation index also fluctuated in an average of 67 sfu and AE index fluctuated with a maximum value of about 658 nT. There was no magnetic storm recorded during this period, although the minimum value of Dst was -29 nT. The particle dynamic was also not very significant. We also find that there was no particles precipitation toward the equatorial region.

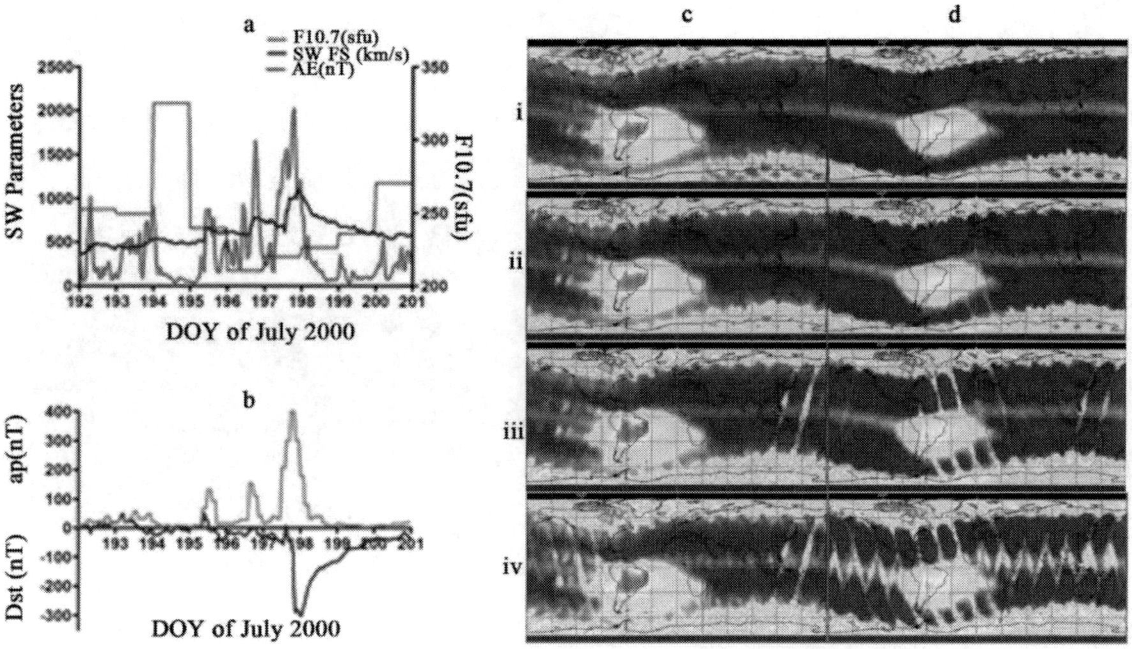

Figure 3. (a) the space weather parameters for July 2000 and (b) the geomagnetic indexes at this period, and the particle distribution for (c) mep0e1 and (d) mep0p2. The i, ii, iii, and iv are represent the mep0e1 and mep0p2 for 13, 14, 15 and 16 July 2000, respectively. The flux distribution unit in Fig. 3c-d is in particle/cm^2 sec str.

4. Summary and Future Work

From a point of view on the potential hazard of space radiation on LEO satellites, we found that three major space radiations have been identified. However, our study was focused on the trapped particle impacts because they have a major role for LEO satellite anomaly rather than other space radiation components. From Figs. 3 and 4, we found that the distributions of particles flux are quite stable at the year of solar maximum and minimum. We also examined the event of a satellite anomaly for those periods, with Astro-D and Orbcomm-1 satellites by combining the space weather and geomagnetic parameters as well as the flux distribution map of particle for analyzing those anomaly events. Our analysis found significant irregularity during the Astro-D anomaly event. This is easy to understand that when Astro-D anomaly happened, the major solar storm that accompanies with X class solar flare and halo CME was correlated with Bastille Day event on 14 July 2000. As for the incidence of Orbcomm-1 found a peculiarities factor, which must still be analyzed because the entire space weather and geomagnetic parameters as well as the distribution of particles when anomalies occur was in a normal state. We estimate that the anomaly of Orbcomm-1 is probably influenced by the length of exposure time of trapped particle in orbit. To deal with this problem in a subsequent study, we suggest calculating the radiation dose absorbed by the satellite during the mission before launch. We also intend to perform in-depth studies of all LEO satellite anomaly events by tracking the location and time of anomalies. By this step, we hope that we can provide an appropriate shielding, material, and mitigation for LEO satellites.

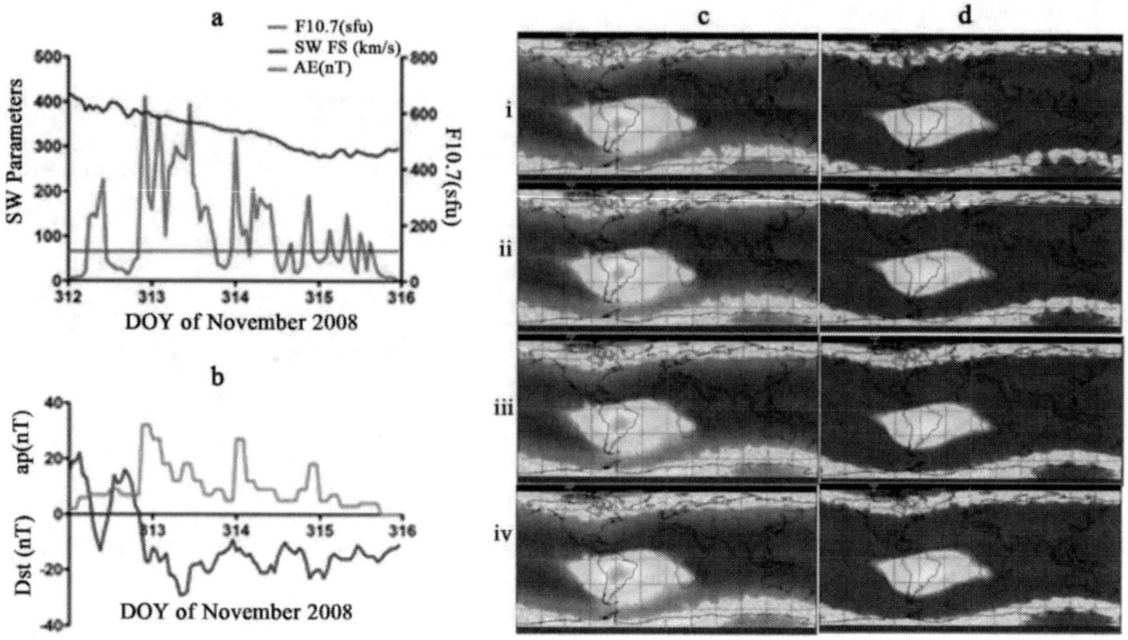

Figure 4. (a) the space weather parameters for November 2008 and (b) the geomagnetic indexes at this period, and the particle distribution for (c) mep0e1 and (d) mep0p2. The i, ii, iii, and iv are represent the mep0e1 and mep0p2 for 8, 9, 10 and 11 November 2008, respectively. The flux distribution unit in Fig. 3c-d is in particle/cm^2 sec str.

Acknowledgments
This work is supported by the Ministry of Science, Technology and Innovation (MOSTI) Malaysia under Science Fund 06-01-02-SF0808 grant and Ministry of Education Malaysia (MoE) under grant FRGS/2/2013/SG03/UKM/02/3. The author thanks to Nizam Ahmad for his help in analyzing the satellite anomaly.

References

[1] Vainio R, Desorgher L, Heynderickx D, Storini M, Fluckiger E, Horne R B, Kovaltsov G A, Kudela K, Laurenza M, Lawlor S M, Rothkaehl H and Usoskin I G 2009 *Space Sci. Rev.* **147** 187
[2] Benton E R and Benton E V 2001 *Nucl. Instrum. Meth. Phys. Res. B* **184** 255
[3] Blandford R and Eichler D 1987 *Phys. Rep.* **154** 1
[4] Asikainen T and Mursula K 2008 *J. Atmos. Sol-Terr. Phys.* **70** 532
[5] Koshiiishi H and Matsumoto H 2013 *J. Atmos. Sol-Terr. Phys.* **99** 129
[6] Hargreaves J K 1992 Cambridge University Press, Cambridge
[7] Barth J L 2004 ed Daglis I A (Dordrecht: Kluwer Academic Publisher) **7** pp 123
[8] Daly E J 2004 ed Daglis I A (Dordrecht: Kluwer Academic Publisher) **5** pp 91
[9] Evans D S and Greer M S 2006 Boulder, CO: NOAA NESDIS NGDC
[10] Tadokoro H, Tsuchiya F, Miyoshi Y, Misawa H, Morioka A and Evans D S 2007 *Ann. Geophys.* **25** 1359

Journal of Physics: Conference Series **539** (2014) 012024 doi:10.1088/1742-6596/539/1/012024

Stochastic effect on thermally magnetization reversal in Pico second ordering process

W. Sadnawanto and Budi Purnama
Physics Department of Post Graduate Program Sebelas Maret University,
Jl. Ir. SUtami 36A Kentingan Jebres Surakarta 57126, INDONESIA

E-mail: bpurnama@mipa.uns.ac.id

Abstract. In this study a magnetization reversal dynamics observed in the heat assisted magnetization reversal of CoPtCr perpendicularly magnetized material in the order of a pico second ordering time. Observation of heat assisted magnetization reversal have been using micromagnetic simulation solved by the Landau-Lifshitz-Gilbert equation. The magnetic dot size of the simulation was 50 nm × 50 nm × 20 nm. The perpendicularly anisotropy constant CoPtCr was 2 × 106 erg/cc. Micro-magnetic simulations was carried out systematically by providing pulse-external field with varying pulse duration of 25 ps to 1 ns. The decrease of the minimum field required for reversal observed in this simulation until it reached 90% for 125 ps cooled ordering time. The results also was showed that this sequence simulation running had a zero probability of switching at zero fields. It was indicated that the heat stochastic effect dominated in the mechanism reduced of the field was required for along to the magnetic field direction.

1. Introduction

Recently, heat assisted magnetic recording (HAMR) is believed as crucial keys to realize high density recording. Here, high anisotropy magnetic perpendicularly as a storage cell memory due to unlimited down scale to nano-meter order lateral dimension. Recording mechanism is carried out by apply heating pulse (nano second order or less) so that temperature rise limit to Curie temperature. When cooling process near room temperature, magnetic field is applied and data save at room temperature. Several studies have reported the final switching distribution [1], localized-heating [2-4] or distribution of barrier energy [5] to explain a low magnetic field which is required to along field direction. Randomly magnetized state due to heating process is considered as a main factor of low magnetic field required for switching. Our previous study show that a configuration of magnetic state when zero field cooling defines the switching field [6]. Gilbert damping also other factor contributes to randomly magnetized state. This paper using micromagnetic simulation by solved the Landau-Lifshitz-Gilbert equation discusses about contribution for both stochastic thermal effect and Gilbert damping in lowering magnetic field required for switching of HAMR.

2. Numerical Methods

In this article, temporal evolution of ordering magnetic state along to magnetic field during cooling process has been investigated. Randomly initial magnetized state is considered for all calculations because temperature increases to near the Curie temperature T_C after heating. The thermal fluctuation effect is taken into account in the Landau-Lifshit-Gilbert (LLG) equation by using fluctuation dissipation effect [7]. In order to calculate the probability of magnetic state along to field direction at given magnetic writing-field amplitude, simulation was performed for 50 different series of random field. In this paper perpendicularly anisotropy magnetic of CoPtCr is considered as a storage cell memory. The perpendicularly anisotropy constant K was 2×10^6 erg/cc and $4\pi M_S$ was 3.75 kG. The magnetic dot size was 50 nm × 50 nm × 20 nm. Micro-magnetic simulations were carried out systematically by providing pulse-external field with varying pulse duration of 25 ps to 1 ns. The second-power temperature dependence with the reduced magnetization was assumed for both the exchange stiffness constant, A and the perpendicular crystalline anisotropy, K [8]. The moderate T_C of 432.9 K was used the whole simulation. The field cooling process start form T_C and the temperature are linearly decreased for ordering cooling-time i.e. from 25 ps to 1 ns.

3. Results and Discussion

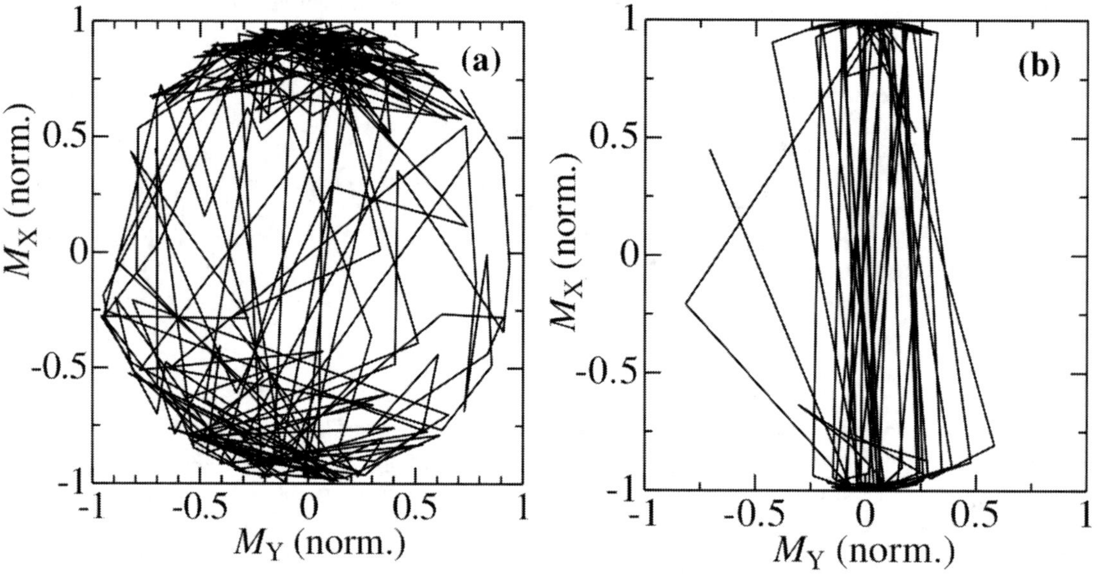

Figure 1 trajectory of randomly magnetized state of nano magnetic particle in M_Y and M_X component under zero-field-ordering cooling from Curie temperature to room temperature for 500 pico second duration calculated for different Gilbert damping constant (a) $\alpha = 0.03$ and (b) $\alpha = 0.3$.

Comparison between two trajectory of randomly magnetized state of nano dot magnetic under zero-field-ordering cooling for 500 ps duration for Gilbert damping constant (a) $\alpha = 0.03$ and (b) $\alpha = 0.3$ as shown in Figure 1. The mean of magnetized state in x-axis was zero. When the randomly magnetized state is considered for lowering switching field of nano magnetic so that nano magnetic with $\alpha = 0.03$ much easy to random than other case. However the sequence simulation showed contrary results as discussed in the later.

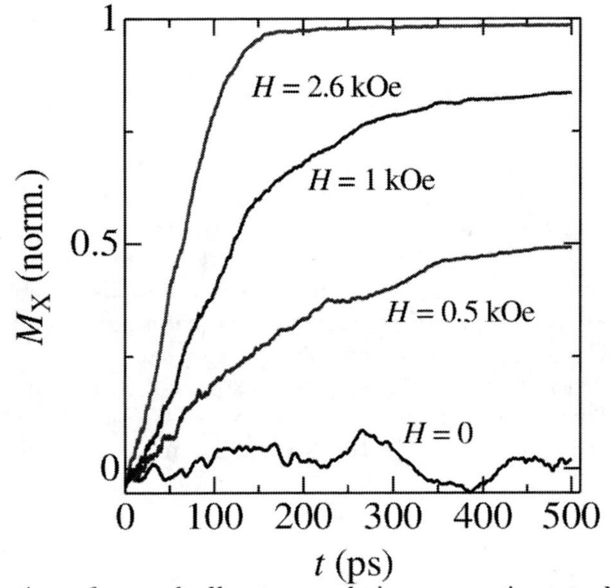

Figure 2 modification of a gradually steps ordering magnetic state M_X (norm.) under a field-ordering-cooling calculated for $K = 2 \times 10^6$ erg/cc, $4\pi M_S$ was 3.75 kG and $\alpha = 0.3$.

Figure 2 show modification of a gradually steps ordering magnetic state M_X (norm.) under a field-ordering-cooling. When $H = 0$, the magnetic nanodot tends to relax into a multi domain configuration within result M_X (norm.) = 0. In order with increasing of magnetic field H, there are not observed the inversed ordering state M_X (norm.). When simulation run for H = 2.6 kOe, heat assisted magnetic recording (HAMR) realized domain nucleation formed of very fast domain wall propagation.

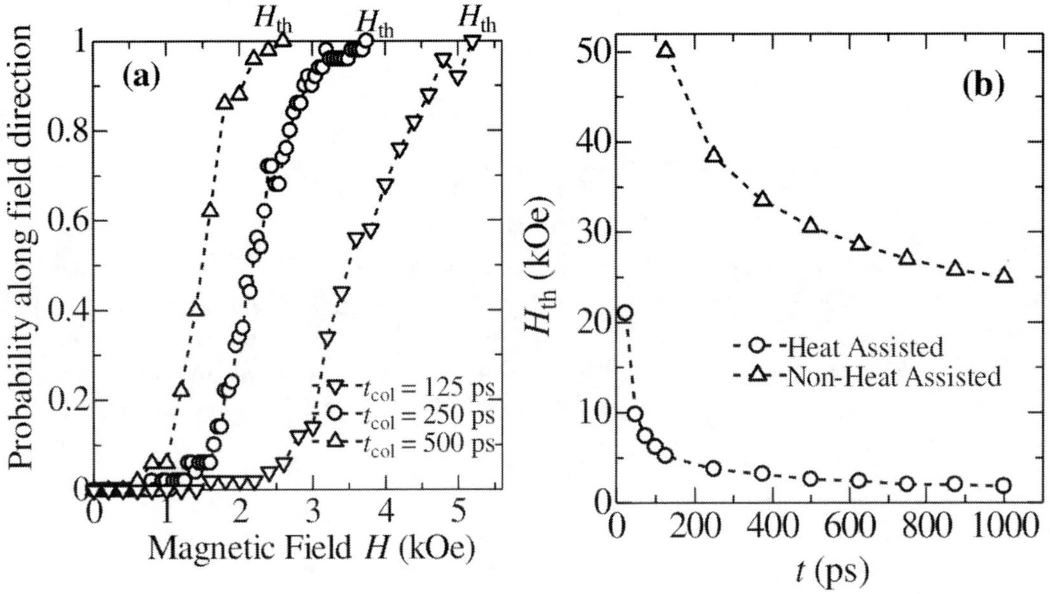

Figure 3. Ordering probability along field direction as a function of the magnetic field H evaluated for sequence cooling time from t = 25 ps to 1 ns (Fig 3.a), and a minimum field required for along field direction H_{th} as a function of t (Fig 3.b).

Figure 3(a) show the ordering probability as a function of the magnetic field calculated for various time ranged from 25 ps to 1 ns. When the ordering magnetic state M_X (norm.) exceeds 0.85, it is found that the inversed domain is completely annihilated. So that, to evaluate the ordering probability of magnetic state, the ordering state is therefore, judged if the (norm.) exceeds 0.85 or not. As shown in Figure 3, the ordering probability gradually increases with the H and reaches to 1 at the threshold field H_{th} which corresponds to the minimum field which is required for thermally assisted magnetic recording (HAMR). For zero magnetic fields, the ordering probabilities have same for all case i.e. zero probability. It can be reasoned when multi domain configuration realize. When the magnetic field increase the domain wall spread to whole magnetic dot within results single domain configuration. Here various times, the ordering probability under zero fields cooling is independence with the time. As shown at the Figure 3(b), the H_{th} decrease over time increase and it become constant at t = 500 ps. It means that the same mechanism for ordering magnetized state occurs when the time much larger than 500 ps. Comparisons two switching field exclude thermally magnetization effect also is depicted at the Figure 3(b). The decrease of the minimum field required for reversal (H_{th}) reaches 90% for 125 ps cooled ordering time. It is indicated that thermally stochastic effect dominated in the lowering the H_{th}.

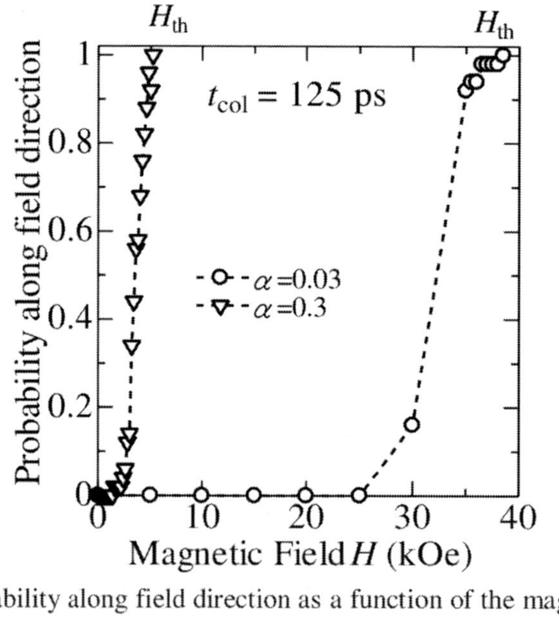

Figure 4 Ordering probability along field direction as a function of the magnetic field H evaluated for cooling time t = 125 ps with different damping Gilbert factor (a) α = 0.03 and (b) α = 0.3.

Figure 4 shows other expression of randomly magnetized state effect which is generated by Gilbert factor on heat assisted magnetic recording (HAMR). As discussed at the previous paragraph that the magnetic dot particle owing low Gilbert damping has high probability to obtain randomly magnetic state. However it compensates by high value of the H_{th} too. In case α = 0.03, the field H_{th} of 38.5 kOe is required to ordering magnetic state occurrences. In the contrary for α = 0.3, it is only 5.2 kOe (less than 1/7) of magnetic field required for 50 sequence simulated successfully switching along to field direction. The value is still in the range of practical application for recording writing technology.

In conclusion, the lowering magnetic field required for ordering along to field direction had clarified by micro-magnetic simulation studied in this paper. Stochastic effects seem to be dominant in the lowering mechanism of HAMR than randomly magnetized state affected by Gilbert constant. Finally the decreased of the minimum field required for reversal (H_{th}) reaches 90% for 125 ps cooled ordering time was obtained in this study.

References

[1] Victora RH and Huang P-W 2013 *IEEE Trans.Magn.* **49** 751
[2] Huang L, Stipe B, Staffaroni M, Juang J-Y, Hirano T, Schreck E and Huang F-Y 2013 *IEEE Trans. Magn.* **49** 2565
[3] Kilic U, Finocchio G, Hauet T, Florez SH, Aktas G and Ozatay O 2012 *Appl.Phys.Lett.* **101** 252407
[4] Li D, Staffaroni M, Schreck E and Stipe B 2013 *IEEE Trans.Magn.* **49** 3576
[5] Ozatay O, Hauet T, Florez SH, Katine JA, Moser A, Thiele J-U, Folks L and Terris BD 2009 *Appl.Phys.Lett.* **95** 172502
[6] Budi Purnama, Nozaki Y and Matsuyama K 2007 *JMMM* 310 2683
[7] Brown Jr WF 1963 *Phys.Rev.* **130** 1677
[8] Mansuripur M and Connel GAN 1984 *J.Appl.Phys* **55** 3049

Journal of Physics: Conference Series **539** (2014) 012025 doi:10.1088/1742-6596/539/1/012025

Space Weather Effects on Microelectronics Devices around the LEO Spacecraft Environments

W Suparta

Space Science Centre (ANGKASA), Institute of Climate Change, Universiti Kebangsaan Malaysia, 43600 Bangi, Selangor Darul Ehsan, Malaysia.

E-mail: wayan@ukm.edu.my

Abstract. Orbiting spacecraft will always be affected by the space environment through their interactions. The main trigger depends on the variability of the radiation environment around the spacecraft such as galactic cosmic rays, solar particle events and the trapped particles. In the low-Earth orbit (LEO), the trapped particles in the radiation belts play a crucial role in affecting the spacecraft in any orbit. The common effects of radiation on spacecraft are charging where some spacecraft components accumulate electric charge from their environment which in turn damaged the spacecraft systems. In addition, space weather will create a stream of charged particles that are associated with solar flares and coronal mass ejections. In this paper, we study the effects of solar radiation on a low orbiting spacecraft (LEO) that would be applied to the equatorial region. Radiation effects in microelectronic components are also discussed. These effects in the radiation environment on an LEO spacecraft are very important in design and testing to mitigate and protect the system from serious damages.

1. Introduction

The failure of Malaysian RazakSat-1 in orbit gives us a stern warning that we need to accommodate the impact of the trapped particles near the spacecraft in the development of a Low-Earth Orbit (LEO)-Equatorial. RazakSat-1 is the second remote sensing near equatorial orbit (NEqO) satellite launched on 14 July 2009 after TiungSat-1. The satellite was designed to operate with a low inclination orbit (9 degrees) at altitude of 685 km, which intended to provide greatly increased coverage of Malaysia with a mission lifetime of 3 years. Unfortunately, they were confronted with an anomaly and terminated in their orbit on 30 August 2010. The cause of the failure is still a mystery. However, we expect that the equatorial orbit makes it trapped in the particle radiation hazards of the South Atlantic Anomaly (SAA). SAA is a part of the inner radiation belt that due to the offset of Earth' magnetic poles and geographic poles lead close to the Earth's atmosphere. SAA is centered in Brazil sky lies from -90° to +40° of longitude and from 0° to -50° of latitude. SAAs are mainly contained of charged particles in several levels of energy up to 10^2 keV for electrons and 10^2 MeV for protons. These particles can harm spacecraft that passed through, especially for LEO orbit [1, 2].

The development of LEO-Equatorial orbit satellite is very important for Equatorial countries, which will increase daily coverage the region than the polar satellite. The challenge is faces more radiation hazard impact rather than the polar satellite because it more frequent passes SAA, and consequently, the exposure time of the trapped particle radiation is longer than the polar satellite. Some sophisticated space weather hazard information system, such as SPENVIS [3] has been developed to simulate space environment hazard. Horne et al. [4] studied the forecasting of high-energy trapped particle in the outer radiation belt area using SPACECAST. Particularly in the inner of radiation trapped particle, models based on physics-magneto hydrodynamic (e.g., [5]) and statistical (e.g. [6]) framework has been employed. We have employed the spatio-temporal hierarchical Bayesian (STHB) model to preliminary forecast the daily trapped particle over the SAA region [7] by applying a Kriging interpolation technique to make the contour distribution map of particle flux in the geographic coordinate [8] and the result was compared with NOAA map.

The purpose of this paper is attempted to study the impact of space radiation hazard in an LEO-Equatorial orbit environment. Due to advantages development of spacecraft in these environments, the radiation damage, total dose and radiation flux, and kind of device loaded should be clearly quantified. As in [9], the major radiation damage can be come from (a) *cosmic rays* in all directions, which majority consisting of protons; (b) *solar particle events* come from the sun direction which consist of a large flux of high-energy (several GeV) protons and heavy ions; and (c) *Van Allen radiation belts* contain electrons (up to ~10 MeV) and protons (up to 100 MeV) trapped in the geomagnetic field. The position of the damage source above poses a concern for satellites, especially a spacecraft that employs complex modern electronics. Studying the effects of radiation environment on LEO spacecraft is very important to mitigate and to protect of our spacecraft from serious damages.

2. Variability of Solar Activity

Figure 1 gives an overview of the solar data for 23rd solar cycle with period of 12 years and 6 months. The variability of solar activity is represented by the sunspot number (SSN), radio solar flux ($F_{10.7}$), solar wind and solar energetic particles (galactic cosmic rays - GCRs). Data for SSN and $F_{10.7}$ on daily basis collected from NOAA - Space Weather Prediction Center (SWPC) website are covered from May 1996 to November 2008, is presented in the first panel. The solar wind proton in the second panel (ACE/SWEPAM) consisting of density (**B**), temperature (**C**) and its speed (**D**) is presented due to it a stream of charged particles released from the Sun. In Figures 1E-F show the GCRs from ACE/CRIS with seven level energies for Carbon (C), Nitrogen (N) and Oxygen (O) elements, and the last panel shows the flux of solar energetic particles (SEPs) from ACE/SIS with a similar element as Figures 1B-D, but we choose for energy level (MeV/nuc) of C_54.3-76.3, N_59.2-83.3 and O_63.8-89. The solar wind, cosmic rays and SEP data from 1 September 1997 to 30 November 2008 are collected from the Advanced Composition Explorer (ACE) satellites (http://www.srl.caltech.edu/ACE/). All the data collected on daily basis is processed into a 27-day average to monitor the trend during one solar cycle.

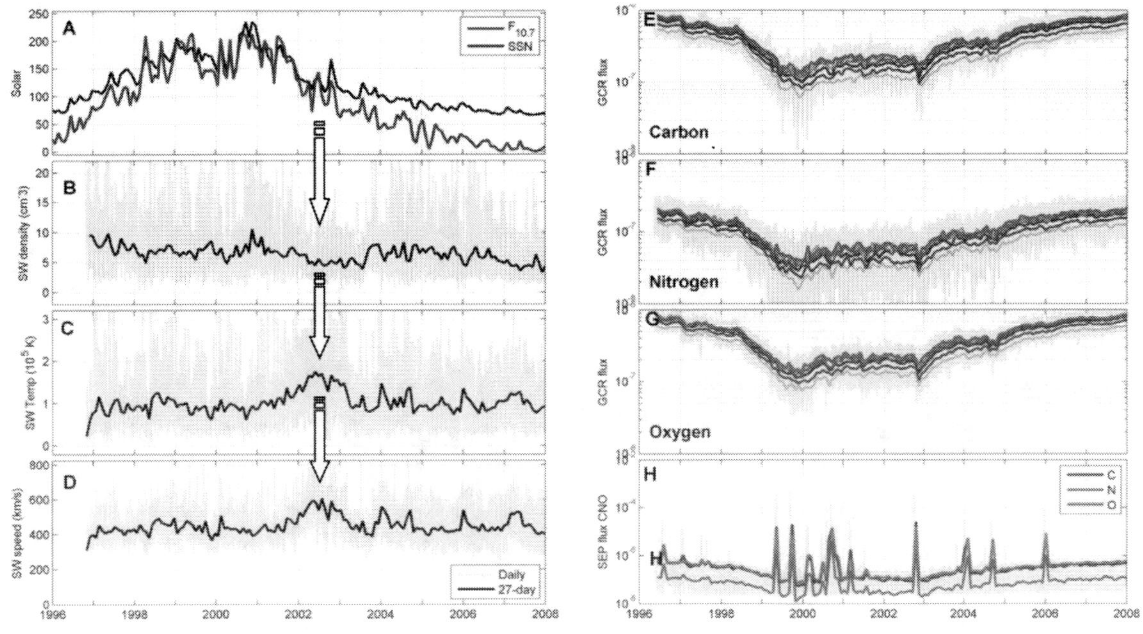

Figure 1. The variability of solar energetic particles during the 23rd solar cycle from May 1996 to November 2008. A down arrow in the left of figure shows the geomagnetic storm on October 2003. The grey background shows the daily average and otherwise is a 27-day average.

As shown in Figure 1, the solar flux maximum was occurred around November 2001. However, one interesting event as indicated by down arrows is Superstorm October 2003. During this event, SSN reached more than 200 and the high-speed stream in the solar wind corresponds to the temperature and density, which indicate that how deep it will penetrate into the Earth's atmosphere. This impacted density of solar wind at the 27-day average, which opposite to that of temperature. In the right panel of Figure 1 clarified that the galactic cosmic rays was opposite to that of SSN/$F_{10.7}$. Although some noises recorded in Figure 1H, the trend of SEP was similar to that of GCR, which both for N are observed 25% lower than C and O. In general, the variability of solar energetic particles that generated space radiation is very crucial to the object in space (e.g., spacecraft). This radiation triggered by GCRs originated from the intergalactic medium, solar particle events (SPEs) which consisted of high fluxes of charged particles, primarily protons associated with solar flares (SFs) and coronal mass ejections (CMEs). One effect of this radiation on spacecraft is charging. Therefore, a specific previous of solar storm in October/November 2003 is presented once to document what has been missed to monitor space weather effects on a spacecraft.

3. Space Weather Phenomena Impact on Spacecrafts

One of the space weather phenomena is a geomagnetic storm. These phenomena can softly or harder damage electronics onboard spacecraft. An example case of space weather associated with geomagnetic storm variation during the Superstorm 2003 (known as the Halloween storm) is presented in Figure 2. The data on hourly basis are from 22 October to 7 November 2003, which covers the period of Superstorm from 29 to 31 October of the year. From top to bottom the panels, the solar activity is characterized by the South (B_z) component of interplanetary magnetic field (IMF B_z), the solar winds (velocity, density and temperature for protons) and SEPs (Ion, Proton and Electron).

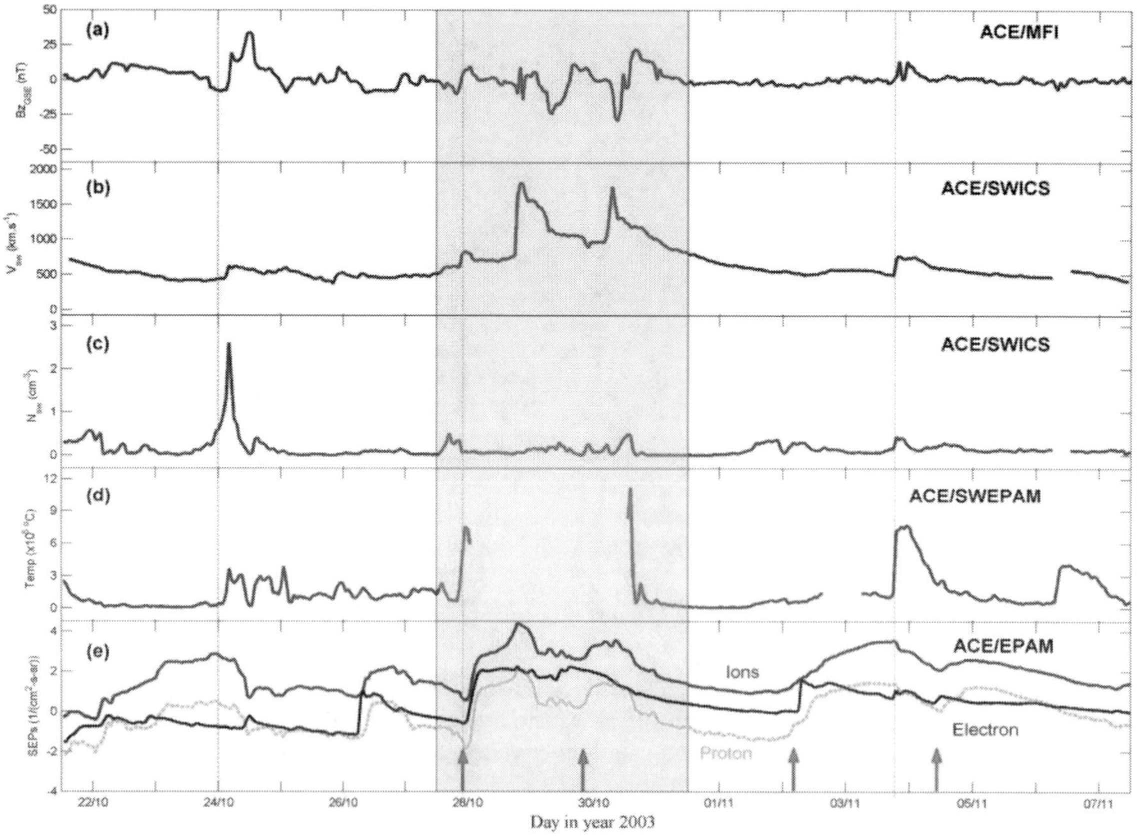

Figure 2. Variability of IMF Bz, solar wind and SEPs during a geomagnetic storm of October 2003.

As shown in the figure, the grey background highlighted are the beginning and ending of the occurrence of the Superstorm. The storm has two distinct episodes, preceded by a high solar activity on solar wind speed (see Figure 2b). The first storm episode peaked by a shock at ~06:00 UT on 29 October with solar wind speed > 1,850 km/s followed by the first shock of IMF B_z minima at ~01:00 UT with value ~-25 nT. The second storm episode was occurred at ~18:00 UT on 30 October with a speed > 1,710 km/s, followed by second IMF B_z minima at ~23:00 UT with value ~-30 nT. During this event, the solar wind density was recorded a low with small ripples and solar wind temperature was missing due to the tremendous of event. However, solar energetic particles were increased together, as indicated by an up arrow in the figure. A complete of explanation of geomagnetic storm on October/November 20013 of this work can be found in Suparta et al. [10] and Suparta [11]. In general, the high plasma densities and temperatures are associated with the interaction of the fast solar wind plasma and magnetic fields (ICMEs) with the slower upstream plasma [12]. More specifically on solar-energetic particles that included in, is presented in Figure 3.

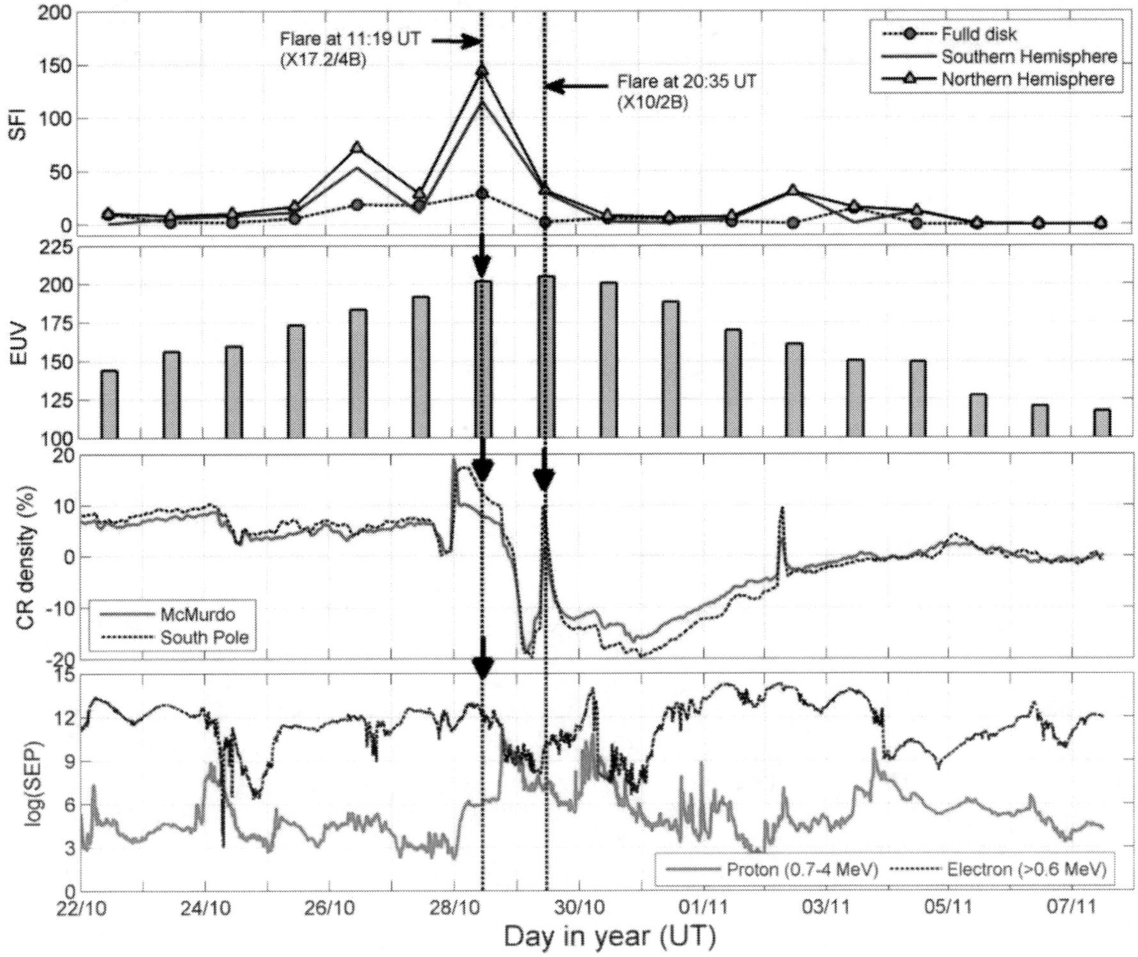

Figure 3. Solar energetic activity and PWV variation observed from 22 October to 7 November of 2003. The vertical dashed line shows the time occurrence of flare activity (X17.2/4B and X10/2B), which shows the shock time wave of both flares, and down arrows shows the association between solar activity and the upper atmosphere parameters. The original figure was reproduced from the paper of Suparta [11].

Since solar particles have unimpeded access to a spacecraft outside the magnetosphere, we present the variation of solar flare index (SFI), extreme ultraviolet (EUV), solar cosmic ray density (SCR) and solar energetic particle variations during the period of events. The peak of SFI for Southern Hemisphere region was measured maximum on 28 October 2003 associated with X17.2/4B flare. Solar full disk is a summation of Southern Hemisphere and Northern Hemisphere regions. The solar EUV (λ = 10-120 nm) radiation affecting the upper ionosphere and is higher during the intense of solar flares. The percentage of cosmic ray (CR) density from ground-based neutron monitors at McMurdo and South Pole stations in Antarctica showed a jump abruptly and the higher fluxes of CR nuclei was modulated by a strong shock wave (SW). The last panel of Figure 3 shows the natural logarithmic of SEP for electrons and protons (obtained from NOAA/GOES), which showed a significa nt enhancement of proton fluxes in both energy intervals associated with SW. In general, the sudden increase of proton peak is coincided with sudden storm commencement and X28 flare activity. This impulse when interact with Earth's magnetic field produce changes in the near-Earth space environment. In addition, solar flares increased the radiation levels in the atmosphere due to the increase in trapped particles in the Van Allen belts. Finally, the particle radiation coming from space, degrade the capability of spacecraft systems, such as microelectronics systems which will be presented in the summary and discussion section.

4. Summary and Discussion

As a brief presented in Section 2 and Section 3, we can see that the sun emits energy that drives changes in the interplanetary environment. The changes in the vicinity of the Earth's magnetosphere modulated by the sun during active and inactive phases of a solar cycle are accelerated by solar flares (SFs) and coronal mass ejections (CMEs). This solar particle event which consists of both protons and heavier ions with different composition will affect inclination and altitude of the orbits. In addition to the solar wind plasma, interplanetary space contains high-energy charged particle called GCRs. During a solar blasting event, solar energetic particles produce significant variation in the CR density. Specifically, spacecrafts are exposed from low to high latitudes, which including particle radiation, ultraviolet irradiation, neutral particles and space debris. Interaction of high-energy ionizing radiation with semiconductor materials can cause degradation, interference or malfunctions of spacecraft systems. Therefore, radiation damage to microelectronic circuits and devices will particularly address in this section. The basic mechanism of how the space radiation environment creates charged tracks in semiconductor electronics is briefly presented.

An important effect occurred from interaction of semiconductor material and high-energy ionizing radiation is so-called the single event effect (SEE). SEE's are caused by high-energy protons and cosmic rays (heavy ion). Heavy ions in the form of electron-hole pairs cause a circuit failure due to plasma track is collected in the sensitive circuit node. The collected at a junction or *p-n* contact is greater than the charge carrying in the threshold component. The most SEE errors are occurred in the SAA region due to protons are trapped in the Van Allen belts. Below are several types of SEE's and they can be classified into soft or hard errors [9, 13]. Soft errors are nondestructive to the device and normally interrupt the normal operations or halt. Hard errors may be physically destructive to the device with permanent functional effects.

- **Single-event upset (SEU)** is a change of state or transient induced by an ionizing particle such as a cosmic ray or proton in a device. This may occur in digital, analog, and optical components or may have effects in surrounding circuit. Penetration of electrons with energies greater than 10 keV can accumulate charge inside the spacecraft resulting various effects such as SEU.
- **Single-event latch up (SEL)** is a potentially destructive condition involving parasitic circuit elements to a single event. In traditional SEL, the device current may exceed device maximum specification and destroy the device if not current limited.
- **Single-event burnout (SEB)** is highly localized burnout of the drain-source in power MOSFETs. SEB is a destructive condition.

- **Single Event Transient (SET)** – an error in term of a reset or rewriting of the device causes normal device behavior thereafter.
- **Single Hard Error (SHE)** - an SEU which causes a permanent change to the operation of a device. An example is a stuck bit in a memory device.
- **Single Event Gate Rupture (SEGR)** - a single ion induced condition in power MOSFETs which may result in the formation of a conducting path in the gate oxide.
- **Multiple Bit Upset (MBU)** - an event induced by a single energetic particle such as a cosmic ray or proton that causes multiple upsets or transients during its path through a device or system.

Figure 4 shows an example of how high-energy of solar radiation threats the microelectronics devices in a CMOS memory cell. In Figure 4a, ionization effects caused charged particles and lattice will be affected due to ones of the ion energy is too low [9]. The ionization effects are usually transient, creating glitches and soft errors. They change the arrangement of the atoms in the crystal lattice, increasing the number of recombination centers, depleting the minority carriers and affected semiconductor *p-n* junctions. One illustration of the SEU soft error in the common 6 transistor SRAM cell is presented in Figure 4b. In that configuration, T_2 and T_4 are PMOS, and T_1 and T_3 are NMOS, where T_5 and T_6 are the access point for reading and writing cell. When the external charged come in (e.g., GCR), the memory cell will in state '1', T_3 and T_2 will be conducted (ON), and T_1 and T_4 will OFF. Consequently, potential at points Q and $\bar{Q}$ are equals V_{cc} and 0 (ground), respectively. If the ion beat deposit with enough charge into the channel beneath the gate oxide (poly-Si), the T_1 will temporally switch ON and the potential at point Q would suddenly reduce and become close to zero. If T_1 is ON long enough, consequently, low voltage occurs at the gate of T_3 and turns it into OFF. This implies that soft effects will occur in T_1, and then T_2 and T_4 will change their state leading to eventually a bit flip.

As a conclusion, the mitigation of these soft errors in modern microelectronics due to space radiation has become important, not only space electronics but also in general commercial electronics. The radiation hazard during geomagnetic storms or substorms is also important to be taken into account in estimating accurately the impact of space weather because of hot plasma will be injected from the magnetotail into the nightside high-altitude equatorial regions that will affect the LEO-Equatorial inclination.

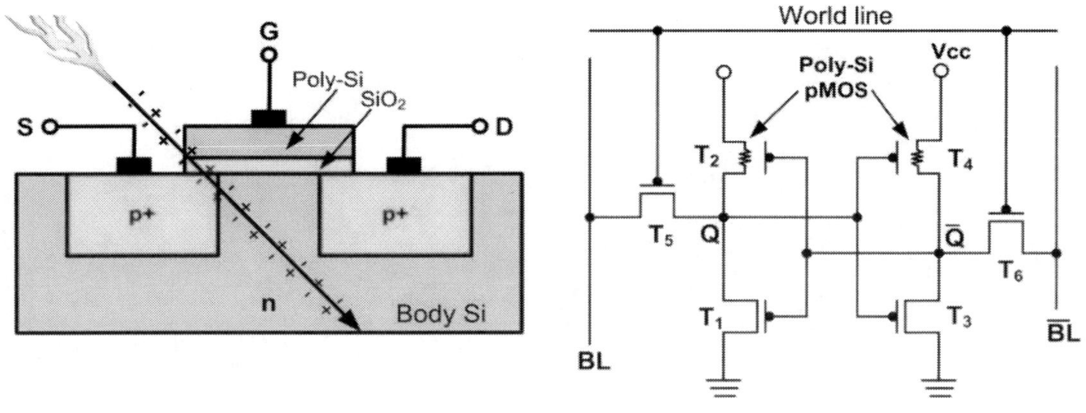

Figure 4. (a) Schematic view of a particle hit in a pMOS transistor and (b) a basic configuration of a 6-transistor SRAM cell

Acknowledgements
This research is supported by the Ministry of Education Malaysia (MoE) under grant FRGS/2/2013/SG02/UKM/02/3. The author would like thank the National Oceanic and Atmospheric

Administration/Space Weather Prediction Center (NOAA/SWPC) for solar indices data; ACE team at GSFC for maintaining the solar wind, GCR and solar energetic particles data; Dr. Tamer Ataç who kindly supplied the solar flare index (SFI) data. Thanks also to Gusrizal my MSc student to help initiating the work.

References

[1] Vainio R, Desorgher L, Heynderickx D, Storini M, Fluckiger E, Horne R B, Kovaltsov G A, Kudela K, Laurenza M, Lawlor S M, Rothkaehl H and Usoskin I G 2009 *Space Sci. Rev.* **147** 187

[2] Casadio S and Arino O 2011 *Adv. Space Res.* **48** 1056

[3] Heynderickx D, Quaghebeur B, Wera J, Daly E J and Evans H D R 2004 *Space Wea.* **2** 10

[4] Horne R B , Glaurent S A, Meredith N P, Boscher D, Maget V, Heynderickx D and Pitchford D 2013 *Space Wea.* **11** 169

[5] Fok M C, Wolf R A, Spiro R W and Moore T E 2001 *J. Geophys. Res.* **106** 8417

[6] Vette J I 1991 *National Space Science Data Center (NSSDC)/ World Data Center (WDC)* p 91.

[7] Suparta W, Gusrizal 2014 *Proc. 2014 Intern. Conf. Sci. Eng. Math. Chem. Phys. (ScieTech)*, Jakarta 13-14 January 2014

[8] Suparta W, Gusrizal, Mohd Alauddin M A and Ahmad N 2013 *Proc. 2013 IEEE Intern. Conf. Space Sci. Comm. (ICONSPACE)*, p 17

[9] Wikipedia 2014 http://en.wikipedia.org/wiki/Radiation_hardening

[10] Suparta W, Abdul Rashid Z A, Mohd Ali M A, Yatim B and Fraser G J 2008 *J. Atmos. Sol. Terr. Phys.* **70** 1419

[11] Suparta W, 2011 *Geographia Technica* **1** 72

[12] Tsurutani B T, Echer E, Shibata K, Verkhoglyadova O P, Mannucci A J, Gonzalez W D, Kozyra J U and Patzold M 2014 *J. Space Wea. Space Clim.* **4** A02

[13] Koselj K 2001 http://merlot.ijs.si/~krizan/belle/svd2/koselj/

Institute of Physics
Dirac House, Temple Back
Bristol BS1 6BE UK

CURRAN ASSOCIATES INC
proceedings.com